**Front
Cover**   Old Man of Hoy, Orkney. Sea stack, 137 m high, of Hoy
          Sandstone resting on pediment of basalt lava (*D1538*)

Plate I   Weisdale Voe, Shetland Mainland, looking south across
          inshore islands to the Clift Hills and Fitful Head
          (*W. Mykura*)   (*Frontispiece*)

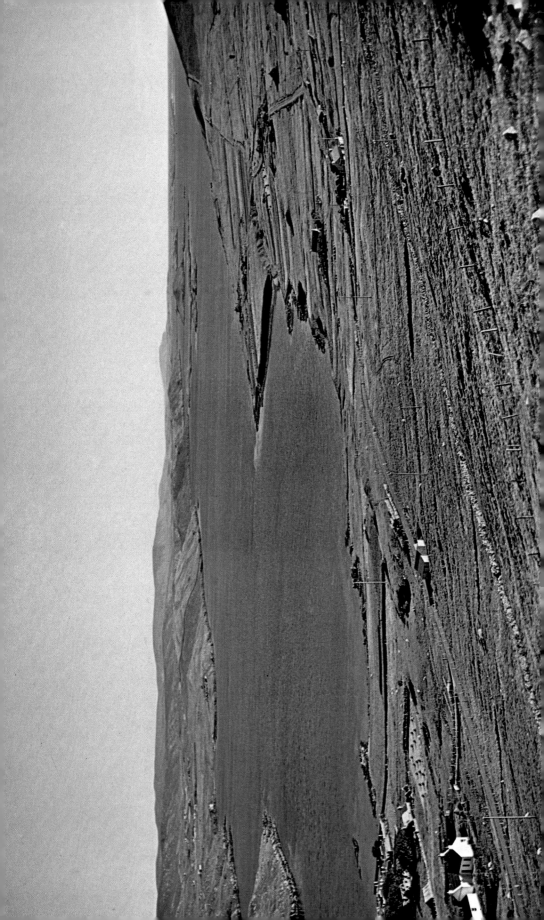

NATURAL ENVIRONMENT RESEARCH COUNCIL
INSTITUTE OF GEOLOGICAL SCIENCES

British Regional Geology

Orkney and Shetland

By W. Mykura, B.Sc., F.R.S.E.

with Contributions by
D. Flinn, B.Sc., Ph.D. and
F. May, B.Sc., Ph.D.

EDINBURGH

HER MAJESTY'S STATIONERY OFFICE

The Institute of Geological Sciences
was formed by the incorporation of the
Geological Survey of Great Britain and the
Museum of Practical Geology with
Overseas Geological Surveys and is a
constituent body of the
Natural Environment Research Council

Crown Copyright 1976

First published 1976

ISBN 0 11 880161 9

FOREWORD

The Orkney and Shetland Islands have not so far been included in any of the British Regional Geology Handbooks because although the area had been completely surveyed, only the maps and memoir for the Orkneys had been published. Post-war revision of the Shetlands by J. Phemister, W. Mykura, F. May and P. A. Sabine has resulted in all but one of the Shetland maps being either published or in press. The remaining Central Shetland map will shortly be in press as a result of the kind co-operation of Dr D. Flinn, who is at present compiling this sheet, based on his own researches. It is unlikely however that full memoir coverage will be available for the Shetland Islands.

Accordingly, although this Handbook covers only a small land area, it has been written in rather more detail than others in the series so that an adequate description of the geology for general purposes should be available, with Chapter 1 forming an initial, brief but comprehensive, summary. The opportunity has also been taken to incorporate modern views on the Old Red Sandstone of Orkney based not only on published work but on the use, by permission, of information from so far unpublished post-graduate theses. Some of the areas to which these theses relate were visited by Mr Mykura in the company of the authors.

Mr Mykura has had a difficult task in the compilation of this account of an area which in its varied geology is certainly one of the most complicated for its size in the United Kingdom. He has of course drawn heavily on the work of the original surveyors. For the Orkney Islands their names are fully acknowledged in the maps and memoir. For the Shetlands they are as follows: J. K. Allan, S. Buchan, D. Haldane, J. Knox, J. Phemister, H. H. Read, T. Robertson, G. V. Wilson. Dr F. May wrote the section on the Spiggie Complex and Dr D. Flinn that on the Brae Complex, while Mr P. Brand checked fossil lists, Mr R. Elliot checked the petrographic material and Messrs A. Christie and J. Pulsford provided most of the photographs.

The assistance of Drs Phemister and May in providing and checking details of their own areas is specifically acknowledged. Of the additional contributions Mr Mykura would like to acknowledge the help given, in Orkney, by N. G. T. Fannin, U. McL. Michie and Miss M. Ridgeway, while in Shetland the ready help and co-operation by Dr Flinn must especially be mentioned. Not only did he make the written contribution referred to above but he provided unpublished details for incorporation dealing with Lunnasting, Whalsay, the Out Skerries, the Graven Complex, Unst and Fetlar, Quaternary Geology and Economic Geology. In addition he read and commented on the whole of the work dealing with Shetland. Dr R. S. Miles named the Shetland fish and Professor W. G. Chaloner the Shetland plants. Dr K. R. Gill read and commented on the Brae Complex section.

Our thanks are due to Aerofilms Ltd. for permission to reproduce Plates IIIA and XVIA; and to the Department of the Environment for Plate XVIB.

The initial editing of the Handbook was carried out by Dr J. D. Peacock, while the Drawing Office staff of IGS Edinburgh produced the numerous diagrams.

Institute of Geological Sciences,   AUSTIN W. WOODLAND
Exhibition Road,   *Director*
South Kensington,
London, SW7 2DE.

CONTENTS

CONTENTS

ILLUSTRATIONS

Figures in Text

Plates

---

[1] Numbers preceded by C or D refer to photographs in the Geological Survey collections.

1. INTRODUCTION

Location

The Northern Isles of Scotland consist of the Shetland and Orkney Islands (Fig. 1). The Shetland Isles lie about 165 km NE of the Scottish mainland and about 340 km W of Bergen in Norway. The island group extends for 109 km from north to south, has a total land area of 1426 km² and consists of over a hundred islands, of which 13 are inhabited. It includes the two outlying islands Foula and Fair Isle, which lie respectively 23 km W and 39 km S of Mainland, the largest island of the group. The Orkney Islands lie about 125 km SW of Shetland and are separated from the mainland of Scotland by the Pentland Firth, which at its narrowest is only 10 km wide. They are formed of about 90 islands and skerries, of which 14 are inhabited. The land area is 956 km² and the islands extend for 80 km from north to south and 47 km from west to east.

Physical Features

The two island groups have strongly contrasting physical features which reflect the difference in their geology. The most striking characteristic of Shetland (Fig. 2) is its north–south elongation and the presence in the eastern half of the island group of smooth north to north-east trending ridges with intervening partly drowned valleys. These features reflect the underlying rock formations which consist of metamorphic rocks and partially recrystallised granites, in both of which the foliation is vertical or steeply inclined and trends north or, in places, north-east. In the western half of Shetland the topography is more rugged and diverse than in the east. This is due to the varied bedrock geology of the area, which includes several large masses of granite and diorite, some belts of metamorphic rock and a large area of highly folded sandstone and lava of Old Red Sandstone age. Slightly less undulating terrain occurs along the south-eastern coastal strip of Mainland and on the adjacent islands as well as on the western seaboard of the peninsulas forming north and west Mainland. These areas are formed of gently inclined sandstones, flagstones and conglomerates of Old Red Sandstone age and, in Papa Stour and Esha Ness, of Old Red Sandstone lavas. The outlying islands Foula and Fair Isle both consist mainly of sandstone which has been eroded into prominent sea cliffs on their western coasts. The cliffs on the west coast of Foula attain a height of 372 m.

The Orkney Islands (Fig. 3), with the exception of western Hoy, have a more subdued topography than Shetland, though in West Mainland, Rousay and Westray hills ranging from 170 m to nearly 275 m in height have small escarpments or less well-defined terrace features on their sides which reflect the alternation of hard and soft layers in the Old Red Sandstone flagstones

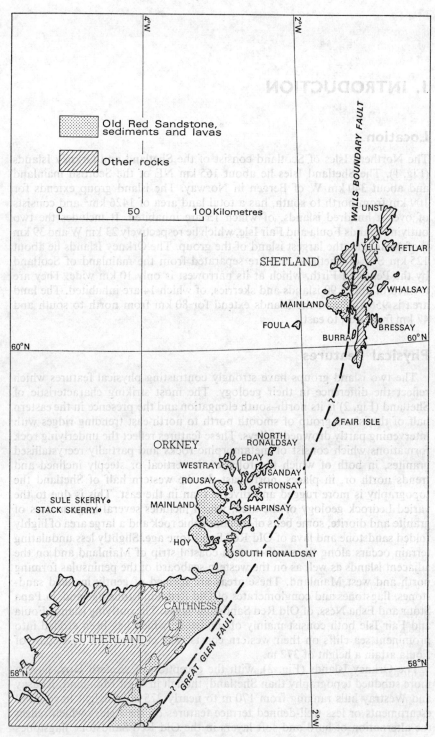

FIG. 1. Location of Orkney and Shetland in relation to the Scottish mainland

that form the greater part of the island group. Considerable portions of the islands of Eday, Sanday, Burray and South Ronaldsay and parts of East Mainland are underlain by thick beds of sandstone, which give rise to ridges and escarpments on Eday and South Ronaldsay. The topography of western Hoy stands out in striking contrast to that of the rest of Orkney. This is because it consists of massive sandstones which form rounded but steep-sided hills up to 477 m high. On the west coast of the island some of the hills are abruptly truncated by magnificent sea cliffs. At St John's Head these reach a height of 335 m.

The Orkney Islands are separated from the Scottish mainland by the east-south-east trending Pentland Firth. Another depression, which trends south-east, separates the Mainland–Rousay–Shapinsay group of islands from the northern isles of Orkney. Smaller south-east trending depressions form the straits between Rousay and Mainland, between Mainland and Hoy and the hollow which is now in part occupied by the lochs of Harray and Stenness and which continues eastwards as Scapa Flow. It has been suggested (Wilson and others 1935, pp. 6–7) that these depressions are the remnants of a late-Tertiary river system which drained to the east-south-east and which was only slightly modified by ice during the Pleistocene Period.

No similar remnants of Tertiary river systems can be recognised in Shetland, though some of the breaks in the north–south trending ridges, such as the Quarff gap 9 km SW of Lerwick, have by some authors been taken as evidence for the existence of 'pre-glacial' or even earlier river valleys.

The ice sheets which covered the islands in the Pleistocene Period and the subsequent rise of sea level around Shetland and Orkney have been major factors in the shaping of the present landscape. In the more rocky parts of Shetland the ice has produced a strongly ice-moulded topography. In Orkney it appears in most areas to have smoothed off and partly obliterated the terrace features formed by the earlier sub-aerial erosion of the flagstones, but there are some hillsides on which ice-gouging may have emphasised these features. The ice has been responsible for the overdeepening of some of the straits between islands and the deepening of some major sea basins such as St Magnus Bay on the west side of Shetland. The ice sheets also scooped out many shallow basins which are now occupied by inland lakes. In Shetland there is evidence that at a fairly late stage in the Pleistocene Period the island group had its own ice-sheet. Even later, after most of the ice had melted from the islands, the higher hills of Shetland Mainland, Foula and Hoy retained small local glaciers. These produced corries and small terminal moraines and the former are prominent features in the landscape of Foula and the north-west corner of Hoy.

The rise of sea level since the last major glaciation is responsible for the drowned landscape topography of Shetland and Orkney. The most obvious signs of this submergence are the many long open inlets or 'voes' along the coast of Shetland, which mark the courses of former valleys. The submergence of the land coupled with the frequent strong winds of the area has also been responsible for the rapid marine erosion along the exposed coasts which has produced the impressive cliffs with their 'geos' (long narrow openings along joints or faults), 'gloups' (caves opening out landward as vertical chimneys or gullies), natural arches and stacks along the exposed shores of both island groups.

The power of the sea during westerly gales is vividly demonstrated by the presence of high level storm beaches. Along certain exposed parts of the coast crescentic mounds composed of large blocks of rock occur on the rocky platform behind a vertical cliff up to 18 m high. Excellent examples of these are to be found in Esha Ness and the Out Skerries, Shetland, and on the west coast of Aikerness in Westray, Orkney. The mounds are up to 5 m high and consist of blocks which have been torn by the waves from the cliff top and from the rocky platform immediately behind.

Along many parts of the coast of these islands one finds long, narrow spits of shingle or sand which are locally known as 'ayres'. Most of these are formed across a shallow bay or across a voe, usually near its landward end, cutting off, either partly or completely, a sheet of water from the sea. Where the spit extends right across the bay the enclosed stretch of fresh water, known as an 'oyce', may eventually silt up to become a fertile stretch of land. Some ayres form tombolos which join off-shore isles to a larger island. Good examples of such tombolos are the three ayres which join the island of Fora Ness to the Mainland of Shetland at Delting, the sand spit linking St Ninian's Isle to south Mainland, Shetland (Plate XVIB) and the ayre at the head of Long Hope, which joins the island of South Walls to Hoy (Orkney).

Summary of Geology

Shetland

Introduction

Shetland consists partly of ancient sedimentary rocks which were metamorphosed and intruded by igneous rocks during the Caledonian Orogeny, and partly of sedimentary and volcanic rocks of Old Red Sandstone (Devonian) age which were laid down and folded during the final phases of that orogeny. It forms a link between the Norwegian, Scottish and East Greenland portions of the Caledonian Orogenic Belt, as they were before disruption by Continental Drift. The understanding of the complex geology of Shetland is thus of considerable importance in any attempt to correlate any of these three remnants of the Caledonian mountain chain. Not only does it provide a link between the stratigraphic and structural units of the metamorphic rocks of the three areas, but it also augments our insight into the palaeogeography of this orogenic belt during Old Red Sandstone times and provides some data about the final localised phases of compression and magma emplacement.

Shetland is divided into two geologically distinct parts, henceforth termed West and East Shetland, by a major north–south trending fault which is termed the Walls Boundary Fault (Fig. 2). It has been suggested that this is the northward continuation of the Great Glen Fault (Flinn 1961b) and like the latter it appears to be a transcurrent fault, in this case with a post-Devonian dextral displacement of 60 to 80 km.

Metamorphic Rocks

West Shetland. Acid and hornblendic gneisses which may be of Precambrian (Lewisian) age form the north-west corner of the Northmaven peninsula and the Ve Skerries, a group of small skerries situated some 6 km

NW of Papa Stour. These rocks (Metamorphic Group A in Fig. 2) may be part of the Caledonian Foreland, separated from the metamorphic rocks of the Caledonian Orogenic Belt to the east by the northward continuation of the Moine Thrust.

East of the 'Lewisian' gneisses a belt of metasediments (Metamorphic Group B in Fig. 2) extends from the northern tip of Mainland southwards along the Walls Boundary Fault to Muckle Roe. Two major rock groups are present in this belt. The older consists of thick steeply inclined bands of impure quartzite, hornblendic gneiss and muscovite-schist. The younger is generally finer-grained and contains two formations with distinctive lithologies: the Greenschist Group and the Calcareous Group. Another strip of metasediments crops out along the north side of the Walls Peninsula and on the islands adjoining this coast. The strike and foliation of the latter rocks is almost east–west and their overall inclination is to the south. They are made up of platy feldspathic muscovite-biotite-gneiss, together with hornblende-schist, tremolite-schist and some calc-silicate-rich limestones. It is probable that these rocks can be equated, at least in part, with the metasediments of Northmaven. The narrow strip of metamorphic rocks along the north-east shore of Foula is similar to the metasediments of the Walls Peninsula.

The metasediments of West Shetland are unlike any of the metamorphic rocks of East Shetland. A study of the deformation history of the rocks suggests that they have been involved in only one orogeny and the potassium-argon dates (435 to 400 m.y.) indicate that this is of Caledonian age; yet it has not been possible to equate them with any members of the Moine–Dalradian sequence of Scotland.

East Shetland. The area east of the Walls Boundary Fault is made up of metamorphic and plutonic rocks, which in the south-east of Mainland are unconformably overlain by Old Red Sandstone sediments. The metamorphic rocks have been grouped into the following three major tectonic units (Fig. 2):

1. The East Mainland Succession.
2. The Quarff Succession.
3. The Unst–Fetlar Nappe Pile.

The *East Mainland Succession* consists of a thick series of north–south trending vertical or steeply dipping metasediments which probably forms a continuous stratigraphic sequence, with the oldest rocks in the west and the youngest in the east. The Succession may be between 22 and 27 km thick and can be split into four major divisions, (Figs. 2 and 6). The most westerly (Yell Sound) division is composed of variably migmatised feldspathic psammites (i.e. metamorphosed sandstones), and has been tentatively correlated with the Moinian of the Scottish mainland. The next (Scatsta) division consists of quartzites, pelitic schists and gneisses, and the third (Whiteness) is made up of flaggy psammite with four thick bands of limestone. The metasediments which form these two divisions are tentatively correlated with the Lower and Middle Dalradian of Scotland. The most easterly (Clift Hills) division consists largely of phyllite with bands of quartzose grit and contains some metamorphosed spilitic lavas. It has been tentatively equated with the Upper and possibly also part of the Middle Dalradian of Scotland (Flinn and

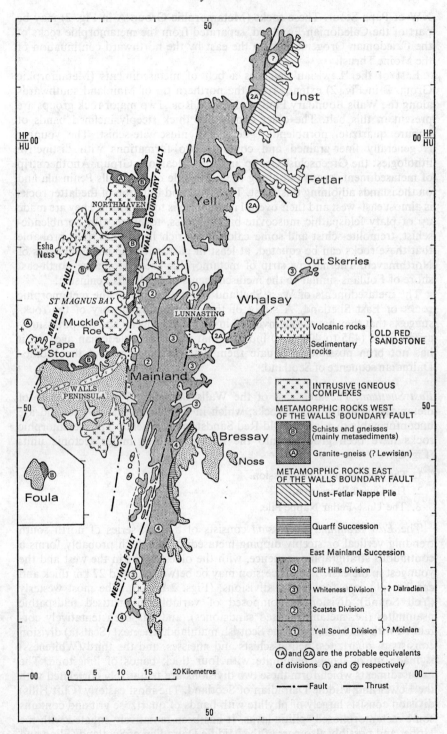

50

HP/HU 00   00 HP/HU

Unst

Fetlar

Yell

NORTHMAVEN

Esha Ness

Out Skerries

St Magnus Bay   Whalsay

Muckle Roe   LUNNASTING

Papa Stour

WALLS PENINSULA

Mainland

Bressay

Noss

Foula

| | Volcanic rocks | OLD RED SANDSTONE |
| | Sedimentary rocks | |

INTRUSIVE IGNEOUS COMPLEXES

METAMORPHIC ROCKS WEST OF THE WALLS BOUNDARY FAULT

(B) Schists and gneisses (mainly metasediments)

(A) Granite-gneiss (? Lewisian)

METAMORPHIC ROCKS EAST OF THE WALLS BOUNDARY FAULT

Unst-Fetlar Nappe Pile

Quarff Succession

East Mainland Succession

(4) Clift Hills Division
(3) Whiteness Division   ? Dalradian
(2) Scatsta Division
(1) Yell Sound Division   ? Moinian

(1A) and (2A) are the probable equivalents of divisions (1) and (2) respectively

—··— Fault   —— Thrust

0   5   10   15   20 Kilometres

50

WALLS BOUNDARY FAULT

MELBY FAULT

NESTING FAULT

FIG. 2. *Geological sketch-map of Shetland*

others 1972, fig. 1) and the Cambro-Ordovician of Norway (Miller and Flinn 1966). The East Mainland Succession contains two great belts of migmatitic gneiss and associated granitic and pegmatitic intrusions. The succession is also intruded by a number of plutonic complexes, which appear to be about 400 million years old.

The island of Yell is made up of a great thickness of migmatised pelites and semipelites with some thick bands of quartzite. These rocks have been tentatively correlated with the lowest (Yell Sound) division of the East Mainland Succession and, by implication, the Moinian of Scotland. The Lunnasting area of Mainland (Plate IV), together with Whalsay, the Out Skerries and the western parts of Unst and Fetlar, are composed of strongly migmatised metamorphic rocks which contain bands of hornblende-gneiss, calcareous schist and limestone. They have been correlated with the two middle divisions of the East Mainland Succession. In Unst and Fetlar these rocks have been locally affected by two phases of retrograde metamorphism, which have become well known through the classic work of H. H. Read (1934c, 1937).

Along part of the east coast of south Mainland the phyllites and spilites of the East Mainland Succession are in tectonic contact, through a mélange of shear-bounded slices, with a suite of permeation gneisses, semipelites and gritty limestones. Though most of these rocks are similar to some members of the East Mainland Succession they are considered to belong to a separate tectonic unit, which is termed the *Quarff Succession*.

The eastern half of Unst and the greater part of Fetlar consist of a number of large tectonic 'blocks' of serpentinite, metagabbro, and, in east Fetlar, a great thickness of deformed conglomerate. These 'blocks', collectively termed the *Unst–Fetlar Nappe Pile*, are separated from each other and from the metamorphic rocks to the west by thrust planes and schuppen-zones composed largely of phyllite, graphite-schist and greenschist. The whole area is thought to form part of two great nappes which were originally parts of an ultrabasic and basic intrusive complex. The schuppen-zone separating the two nappes contains pebbles and debris derived from the erosion of the lower and possibly also the upper nappe. This suggests that the nappes of this area were emplaced at a very high tectonic level, possibly by gravity sliding on the surface of the developing Caledonian mountain chain.

Old Red Sandstone

West Shetland. The Old Red Sandstone of West Shetland falls into two distinct groups which are separated from each other by the north-north-east trending Melby Fault. East of this fault is the immensely thick Lower to Middle Old Red Sandstone outcrop of the *Walls Sandstone*, which forms the greater part of the Walls Peninsula. This can be divided into two litho-stratigraphical units, the *Sandness Formation* and the *Walls Formation*. The former includes the Clousta Volcanic Rocks near its top. Both formations have been affected by two major episodes of folding, the first of which produced a complex east-north-east trending synclinorium and the second a number of tight north to north-east trending folds. There are also several small folded outliers of sandstone and breccia of possible Devonian age in North Roe and on the island of Gruney, 2 km N of the most northerly point of Mainland.

The Old Red Sandstone west of the Melby Fault is termed the *Melby*

B

Formation. It contains two fish-bearing horizons, the *Melby Fish Beds*, which have been correlated with the Sandwick Fish Bed of Orkney. The sediments of the Melby Formation appear to be conformably overlain by basic lavas, tuffs and rhyolites which form the island of Papa Stour and the peninsula of Esha Ness in Northmaven.

The greater part of Foula consists of soft buff, gently inclined sandstones with subordinate siltstones and shales which have yielded plant remains of Devonian aspect.

East Shetland. The sedimentary rocks which crop out along the south-eastern coastal strip of Mainland and on the islands of Bressay and Noss rest unconformably on metamorphic rocks. The surface of unconformity appears to have a high relief and the base of the overlying succession is strongly diachronous. The sequence consists in part of sandstones with lenticular masses of conglomerate and basal breccia and, higher in the succession, of interbedded sandstones and flagstones. The fish remains suggest that the age of these beds is high Middle Devonian (Givetian) and that the highest beds may be equivalent to part of the Upper Old Red Sandstone of the Scottish mainland.

Fair Isle. This island consists of a thick sequence of medium-grained sandstone and pebbly sandstone with subsidiary bands of dolomitic mudstone and shale. Plant remains and fish fragments suggest that the age of the beds may range from Lower to Middle Old Red Sandstone.

Late Caledonian Intrusions

The metamorphic and sedimentary rocks west of the Walls Boundary Fault have been intruded by a series of probably interconnected plutonic complexes composed mainly of granite, diorite and subordinate gabbro which, from radiometric studies, appear to be between 350 and 360 million years old. Great swarms of roughly north–south trending acid, intermediate and basic dykes cut the more northerly of these plutonic complexes and the adjoining country rocks. East of the Walls Boundary Fault the metamorphic rocks are cut by dykes and sills of microdiorite, many of which are slightly affected by shearing and thermal metamorphism, and also spessartite.

The Old Red Sandstone sediments of Bressay and Noss contain two north–south trending belts of steeply inclined strata which are associated with vents and with more irregular masses of non-igneous tuffisitic breccia and tuffisite.

Orkney

Stratigraphy

The Orkney Islands (Fig. 3) consist almost entirely of sedimentary rocks and subordinate lavas and tuffs of Middle and Upper Old Red Sandstone age. A *Basement Complex* composed of metamorphic rocks of Moinian type and Caledonian granites forms a number of small inliers near Yesnaby and Stromness in West Mainland and on the island of Graemsay (Fig. 16). At Yesnaby there is also a group of sandstones and breccias which may be of *Lower Old Red Sandstone* age, and at Warebeth, west of Stromness, purple siltstones and sandstones, possibly of similar age, have been encountered in a borehole.

The *Middle Old Red Sandstone* falls naturally into two major groups. The lower group, comprising the *Stromness Flags* and the *Rousay Flags*, consists largely of 'flagstones' and is made up of rhythmic sequences of thinly bedded and, in part, laminated grey and black carbonate-rich siltstones and silty mudstones alternating with generally thin beds of fine-grained sandstone or sandy siltstone. The flags have yielded well-preserved fossil fish and the Stromness Flags contain the Sandwick Fish Bed, which is considered to be the equivalent of the Achanarras Limestone of Caithness.

The upper group, the *Eday Beds*, comprises the Lower, Middle and Upper Eday Sandstone, three thick sequences of yellow and red sandstone with pebbly lenses, which are separated respectively by the Eday Flags and the Eday Marls. The Eday Flags locally contain a few thin flows of basic lava and some thin beds of tuff.

Beds ascribed to the *Upper Old Red Sandstone* are confined to the island of Hoy, where they form up to 1000 m of red, pink and yellow sandstones with subordinate bands of marl. They are underlain by a variable thickness of basalt lava and tuff which rest on a hummocky surface floored by various members of the Middle Old Red Sandstone sequence.

Structure

The outcrops of the major rock groups and the axial traces of the major folds of Orkney are shown in Fig. 3. Most folds affecting the Old Red Sandstone are very open so that the strata are generally gently inclined, but there are a few folds of limited regional extent which have steeply inclined limbs. Many of the principal folds have a near-northerly trend. The most important of these is the northward-plunging Eday Syncline which determines the structure and physiographic pattern of the northern group of islands. Other well-marked flexures of regional importance are the West Mainland Anticline and the Deerness Syncline. Of equal importance in determining the structural pattern of Orkney and the disposition of the various rock groups are the faults. The largest and most important of these are the North Scapa Fault, the East Scapa Fault and the Brims–Risa Fault (Fig. 3). The latter two are reversed faults for at least part of their course. Smaller faults are numerous and closely spaced, but many of them have relatively small displacements. The faults can be divided into the following three systems according to their trend:

a. East-north-east to north-easterly, e.g. North Scapa Fault and major faults cutting Shapinsay and South Ronaldsay.
b. Northerly (i.e. with trends ranging from north-north-west to north-north-east), e.g. East Scapa Fault, faults in South Ronaldsay and Westray and a suite of sub-parallel step faults in West Mainland.
c. North-westerly. These are less common.

Adjacent faults generally have the same direction of downthrow. Movement on many faults began before the deposition of the Upper Old Red Sandstone and some, such as the North Scapa Fault, were probably active during and possibly even before the deposition of the Middle Old Red Sandstone Eday Beds, whose thickness appears to be greatly increased on their downthrow sides.

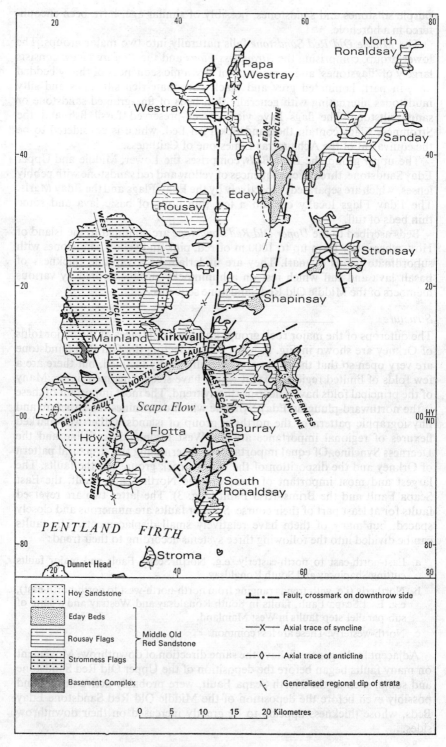

Fig. 3. Geological sketch-map of Orkney

A characteristic feature of the flagstones of both Orkney and Caithness is the presence of narrow belts of intensely folded and contorted sediments. These belts generally have an overall monoclinal pattern (Plate XIB) and a trend which is parallel to the dominant fault system of the area. Some of these tight monoclines contain belts of mineralised breccia.

Minor Intrusions and Volcanic Vents

The earliest intrusive rocks cutting the Old Red Sandstone sediments of Orkney are small masses of teschenitic dolerite which are associated with the alkaline basic lavas and tuffs in the Eday Flags. Most of the dykes and sills of Orkney, however, belong to the suite of late-Carboniferous camptonites, monchiquites and bostonites and they are most numerous in the south and west of Orkney, with camptonites predominating in West Mainland and Rousay and monchiquites in the southern islands.

A number of small volcanic vents have been recorded, principally in Hoy, South Ronaldsay and East Mainland. Some of these are closely connected with monchiquite dykes and a few are partly filled with monchiquitic basalt. Several of the vents contain breccia composed entirely of sedimentary rock and these resemble the breccia vents of East Shetland.

History of Research

Shetland

Early Accounts. The earliest references to the rocks of Shetland are records of the occurrence of useful minerals, ores and semi-precious stones. Thus George Low, who visited the islands in 1774, recorded the presence of talc and 'ironstone' (?serpentine) in Unst, fuller's earth and copper ore in Fetlar and bog iron ore on Vaila. A record of the presence of iron ore in Dunrossness and of copper ore being mined in southern Shetland first appeared in the Statistical Account of Scotland (Sinclair 1793, vol. V, p. 186, and Vol. VII, p. 393). The first attempt at a systematic account of Shetland geology is that of Jameson (1798, 1800), who commented on the dependence of the topography on the strike of the rocks and recognised that the Walls Sandstone is related to the sandstones and conglomerates seen along the coast of south-east Mainland. Jameson described many of the metamorphic minerals of Shetland and all the metallic ores, except chromite, which are known today. Further accounts of Shetland mineralogy and geology were produced by Traill (1806) and Fleming (1811 and *in* Shirreff 1817). These also deal with the working of copper and iron ores in south-east Mainland and the occurrence of copper in Fair Isle.

The most important of the early works on the geology of Shetland are those of Hibbert (1819–22), which include the first geological maps of the islands. Hibbert dealt with the distribution and structural relationship of the Shetland rocks, and also noted the presence of brucite and chromite. He drew attention to the great induration of the Walls Sandstone, which he called 'primitive quartz rock'. Ami Boué's (1820) account of Shetland geology incorporated the observations of the earlier workers, and attempted the first correlation of the metamorphic rocks of Shetland with those of Scotland, equating the gneisses of Mainland, Yell and Unst with the Lewisian of Sutherland and the Hebrides, and the chlorite-schists and clay slates of Shetland with the Dalradian. He recognised the similarity between the sandstones and conglomerates of Shetland and the sedimentary rocks of the Moray Firth–Caithness–Orkney terrain and pointed out that the rocks of Esha Ness and Papa Stour resemble the products of extinct volcanoes. Other early works dealing with Shetland are those of MacCulloch (1821, 1831) and Nicol (1844).

Old Red Sandstone. Fossil plants of Old Red Sandstone age were first discovered at Lerwick and Bressay (Tufnell 1853; Hooker 1853; Murchison 1853, 1859) and these were thought to belong to the Upper Old Red Sandstone. Later Peach and Horne (1879a, *in* Tudor 1883; Geikie 1879) found plants of *Psilophyton* type in the Walls Sandstone and thus proved that this formation is of Old Red Sandstone age. They also established a stratigraphic sequence in the Old Red Sandstone of east Shetland and described the Old Red Sandstone volcanic rocks of west Shetland (Peach and Horne 1879a, 1884). Fossil fish from south-east Mainland were first recorded by Flett (1908) who suggested that the beds of this area may range in age from the Middle to the Upper Old Red Sandstone. Finlay (1926b, 1930) produced comprehensive accounts of the Old Red Sandstone sedimentary, volcanic and plutonic rocks of Shetland and recorded the discovery of the important fish bed at Exnaboe in south-east Mainland. Much additional information was obtained during the mapping of the islands by the Geological Survey which took place between 1929 and 1934, the most important, perhaps, being the discovery by Knox of the Melby Fish 'Bed'. The fauna of this 'bed'* was described by Watson (1934) and shown to be similar to that of the Sandwick Fish Bed of Orkney. More recently Westoll (1937, 1951) has produced a zonal correlation of the fish-bearing east Shetland sequence with the Devonian sequences in the Baltic and East Greenland regions. The account of the Old Red Sandstone in this handbook is largely based on the recent work of Mykura (1972b, Mykura and Phemister 1976).

Post-Orogenic Intrusions and Mineralisation. Investigations into the plutonic and hypabyssal rocks of West Shetland by Peach and Horne (1884) and Finlay (1930) were greatly amplified by the Geological Survey mapping. Detailed petrographic studies of the riebeckite-felsite dykes of Northmaven were made by Phillips (1926) and Phemister (1950). Mykura and Young (1969) and Mykura (1972c) described the scapolite associated with shear zones and basic intrusions in the Sandsting Complex and in Fair Isle. Accounts of the plutonic complexes of Sandsting and Northmaven were produced by Mykura and Phemister (1976) and by Pringle (1970).

Metamorphic rocks. Prior to 1929 the main work on the metamorphic and synorogenic igneous rocks of Shetland was carried out by Heddle (1878, 1901) whose primary task was the description of the minerals and mineral localities throughout the islands. Heddle recorded the presence of many rare minerals, some minerals new to science, and also the first known British occurrence of relatively common minerals such as chloritoid. He also produced detailed accounts of the geology of certain parts of Shetland and discussed the age and possible correlation of the Shetland metamorphic rocks. He

---

* The Melby Fish Bed recorded by Knox is now known to consist of two beds, separated by about 110 m of sediment (p. 8).

suggested that the latter form a group found nowhere else in Scotland, being apparently older and more highly metamorphosed than the conglomerates, schists and gneissose rocks which overlie the Lewisian (i.e. presumably the Moinian) but not so ancient as the Lewisian gneiss.

More recent work on the metamorphic rocks of Shetland commenced with the petrographic studies of serpentines, metagabbros, epidotic granites and andalusite-schists by Phillips (1927–28), and was continued with the mapping of the entire outcrop by the Geological Survey (Summ. Prog. 1930 to 1935). As a result of this survey Read (1933, 1934a, b and c, 1937) published a series of classic papers dealing with the geology and polymetamorphism of Unst, the phases of retrograde metamorphism in the Valla Field Block of north-west Unst, and the development of zoned bodies from ultrabasic intrusions. Further work on the ultrabasic rocks of Unst was carried out by Amin (1952, 1954), on augen gneisses by Fernando (1941) and on chloritoid-schists by Snelling (1957, 1958). Our present knowledge of the stratigraphy, structure and metamorphic history of eastern Shetland is, however, based largely on the work of Flinn, who, in a series of important papers, has described and interpreted the structural and metamorphic history of Delting (1954) and south Shetland (1967a), the nappe structure of Unst and Fetlar (1958, 1959) and the tectonic evolution of the Funzie Conglomerate of Fetlar (1956) and the Muness Phyllite of Unst (1952). In conjunction with Miller (Miller and Flinn 1966) he has produced the first integrated account of the geology of Shetland, in which the major tectonic, metamorphic and igneous events are placed in a time perspective and in which the main groups of metamorphic rocks are tentatively correlated with their possible equivalents on the Scottish main-land and in Scandinavia. May (1970) has established a sequence of meta-morphic, migmatitic and structural events in the rocks of the Scalloway region, and, with Flinn (Flinn and others 1972), he has re-interpreted the stratigraphy of the East Mainland Succession. The metamorphic rocks west of the Walls Boundary fault have been described by Pringle (1970), Phemister (1976) and Mykura (*in* Mykura and Phemister 1976).

Pleistocene and Recent. The first integrated account of the glaciation of Shetland is that of Peach and Horne (1879b), who suggested that the islands were first overridden by Scandinavian ice from the north-east, but that they later nourished their own ice cap. The former presence of Scandinavian ice was confirmed by the presence of an erratic of tönsbergite in south-east Mainland, which was recorded by Finlay (1926a). Robertson (1935) and Flinn (1964, 1967b) have, however, suggested that during the last glacial maximum Scandinavian ice crossed only the extreme south and north of the Shetland Islands and that over the rest of the area a locally-nourished ice cap prevailed. The changes in sea level around Shetland since the last glaciation have been studied by Flinn (1964) and Hoppe (1965), and a floral sequence in the Shetland peats was established by Lewis (1907, 1911). Chapelhowe (1965) recorded the presence of an interglacial peat deposit in north-west Mainland and Mykura has since found a similar bed in west Mainland. These peats have given ^{14}C dates of 35 000 to 40 000 BP and Birks and Ransom (1969) have suggested, on pollen analytical evidence, that they may be of Hoxnian age.

Economic Geology. Records of early workings for copper and iron at and near Sand Lodge and at Garths Ness (south Mainland) appear in many accounts, the most important of which are by Traill (1806), Fleming (*in* Shirreff 1817), Heddle (1880) and Dron (1908). The available data were summarised by Flett (*in* Macgregor and others 1920 and *in* Wilson 1921). A description of the Sand Lodge workings in 1929 was given by O'Dell (1939). The magnetite ore at Clothister Hill near Sullom Voe was discovered by D. Haldane and the exploratory work connected with the deposit has been described by Groves (1952). The history of the chromite workings in Unst has been documented by Sandison (1948). There is also an account of the more recent explorations for this mineral by Rivington (1953). Summaries of the workings in and distribution of non-metallic minerals, such as serpentine, talc and magnesite, have appeared in the economic memoirs of the Geological Survey (*see* Strahan and others 1916; Macgregor and others 1940; Wilson and Phemister 1946), and in reports of the Mineral Resources Panel of the Scottish Council (1954a, b). A detailed assessment of the talc-magnesite deposits near Cunningsburgh was carried out by Bain and others (1971), and the distribution, composition and petrography of the worked and workable limestones of Shetland have been described in Geological Survey memoirs (Robertson and others 1949; Muir and others 1956).

Geophysics and Off-shore Geology. The Institute's regional geophysical survey of the Shetland Islands was carried out by McQuillin and Brooks (1967) and Sheet 16 of the *IGS* 1:250 000 Aeromagnetic Map of Great Britain and Northern Ireland, which includes the sea area around Orkney and Shetland, was published in 1968. Flinn (1969a) has interpreted the data from this and has produced a map showing the possible geology and structure of the areas north of Scotland. Exploratory marine geophysical surveys by Watts and Bott (Bott and Watts 1970; Watts 1971) have further added to our knowledge of the geology of the Shetland seas, and bathymetric investigations by Flinn (1964, 1969b) have contributed to the understanding of possible pre- and post-glacial movements of the land relative to sea level.

Orkney

The earliest geological records of Orkney deal with the presence of lead ore (Speed 1666) and minerals and rocks such as granite, limestone, flagstone and millstone (Sinclair 1795) which were being worked on these islands. Jameson (1800) noted the presence of granitic rocks, overlain by breccia and flagstones near Stromness, and recorded basalt 'veins' (i.e. dykes) at Yesnaby and basalt under sandstone at Shapinsay. He also described the conglomerate ('sandstone breccia') at Hegglie Ber in Sanday. A record of early collections of fossil fish appears in the Proceedings of the Orkney Natural History Society (1905), which tells of about 100 specimens exhibited in 1837 in the Stromness Museum.

A large collection of fossil fish by Professor Traill was identified by Agassiz (1833–43) who established the correlation of the Orkney flagstones with those of Caithness. Localities from which fish were collected are mentioned by Clouston in the 'New Statistical Account of Scotland' (vol. 15, 1845). Widespread interest in the Orkney fish was awakened by Hugh Miller (1849, 1858) who recorded and described several species of fish found by himself and a number of local collectors, and was able to identify forms which also occurred in the flagstones of Caithness. Murchison (1859) noted that the Old Red Sandstone of Orkney could be divided into a lower flagstone series and an upper sandstone series, and Geikie (1879) recognised the unconformity at the base of the Hoy volcanic rocks, and ascribed the Hoy sandstones to the Upper Old Red Sandstone. He also regarded the sandstones of South Ronaldsay as the lateral equivalents of the similar rocks near John o'Groats and concluded that the Old Red Sandstone rocks of Orkney, Caithness, Ross and Cromarty were laid down in a great intermontane basin which he termed 'Lake Orcadie'.

The geology of the northern islands of Orkney was first described by Heddle (1880) who showed that, as on Mainland and South Ronaldsay, the flagstones are conformably overlain by a series of sandstones which are locally conglomeratic. He also noted that the strata forming these islands were folded into a well-marked syncline. Peach and Horne (1880, *in* Tudor 1883) produced the first comprehensive accounts of the geology of Orkney and confirmed most of the conclusions arrived at by Geikie and Heddle. They also described the glaciation of the islands (1880). Our knowledge of the fossil fish of Orkney is largely based on the monograph by Lankester and Traquair (1868–1914), and on a series of papers by Traquair, which have also formed the basis of correlation of the Sandwick Fish Bed with the Achanarras Limestone of Caithness and the fish beds of Cromarty. Flett (1897) produced the first detailed stratigraphy of Orkney, in which he recognised the Rousay Beds as a separate formation. He recognised the presence of the John o' Groats fish fauna in the Eday Beds and also described in detail the lamprophyric minor intrusions of Orkney (Flett 1900). Work on the Stromness basement rocks and the geology of Stronsay was carried out by Steavenson (1928 a and b) and the peat deposits of Mainland and Westray were examined by Erdtman (1924–29).

The Geological Survey mapped the islands between 1927 and 1929 and a descriptive memoir was published in 1935. The officers engaged in this survey were Wilson, Edwards, Jones, Knox and Stephens. The fish remains collected during the survey were determined by Watson and the plants by Lang. Much of the present account is based on this memoir. Additional work on the fish remains has been carried out by Miles and Westoll (1963) and the latter has also correlated the Orkney sequence with the marine Devonian succession (Westoll 1951). More recently Kellock (1969) has re-examined the alkaline igneous rocks associated with the Eday Beds. Fannin (1969, 1970) has carried out a detailed sedimentological study in the Stromness Flags, and Miss Ridgway has similarly investigated the Eday Beds. The *IGS* regional geophysical survey of Orkney was carried out by McQuillin (1968).

Selected References[1]

Finlay 1926b, 1930; Flinn 1954, 1958, 1961b, 1964, 1967a, 1969a; Flinn and others 1972; Heddle 1878; Hibbert 1821; McQuillin 1968; McQuillin and Brooks 1967; May 1970; Miller and Flinn 1966; Mykura 1972a, b; Mykura and Phemister 1976; O'Dell 1939; Peach and Horne 1879b, 1884; *in* Tudor 1883; Phemister 1976 Phemister and others 1950; Read, 1934c, 1937; Watts 1971; Wilson and others 1935.

[1] A work by one author commonly deals with several subjects described separately in the following chapters of this handbook. In order to avoid needless repetition of the full title only abbreviated references are given at the end of each chapter. The full titles can be found in the general Bibliography at the end of this handbook, against the name of the author and date of the publication.

2. METAMORPHIC ROCKS OF SHETLAND: I AREA WEST OF WALLS BOUNDARY FAULT

The metamorphic rocks west of the Walls Boundary Fault form a belt of variable width which extends from North Roe at the extreme northern end of Shetland Mainland southwards along the fault as far as the island of Papa Little [34 61][1], (Fig. 2) and thence westward for 15 km along the northern margin of the Walls Peninsula. There are also a number of outcrops separated from the main belt by the granite or diorite of the Northmaven Igneous Complex. These include the rocks of the Hillswick area and the 'enclaves' in the granite and diorite of Muckle Roe and the Busta Peninsula. Metamorphic rocks also form the Ve Skerries [10 65] and the north-eastern coastal strip of Foula.

North Roe and Northmaven

In North Roe and Northmaven the rocks have been divided into three major series, shown in Fig. 4. Of these the *Western Series* comprises gneisses which may be of Lewisian age and are older than the other two series. While no direct correlation is possible, the rocks of the Ve Skerries are thought to belong to this basement group. The *Fethaland Series* and the *Ollaberry Series* have been definitely distinguished only in the eastern part of the area, but it is likely that the rocks of the Hillswick Peninsula and the inclusions in the Northmaven Igneous Complex can also be referred to these series. Tentative correlations are shown in Fig. 4. The Fethaland and Ollaberry Series consist of sedimentary and volcanic rocks metamorphosed during the Caledonian Orogeny. They are separated from the Western Series by a shear zone, the Wester Keolka Shear Zone, which may be a major thrust comparable to the Moine Thrust of the Scottish mainland. The true relationship of the Fethaland to the Ollaberry series is likewise complicated by shearing along the contact, but it is suggested that the Ollaberry Series is the younger.

In the Walls Peninsula the metamorphic rocks have been divided into four lithostratigraphic groups whose distribution is shown in Fig. 5. The metamorphic rocks of Foula have been likened to one of these groups (p. 23), although no direct correlation is possible. No direct correlations have been made between the metamorphic rocks of the Walls Peninsula and the two eastern series of Northmaven, but the similarities in the metamorphic and tectonic history of the two rock-groups suggest that they share a common tectonic environment and are likely therefore to be of similar age. Although these rocks appear to have been deposited within the Caledonian Geosyncline they cannot be readily correlated with any part of the Dalradian or Moinian assemblages of the Scottish Highlands. Radiometric potassium-argon ages which average 415·5 m.y., obtained from the rocks of the Walls Peninsula,

---

[1] Figures in square brackets are National Grid References.

do not give any clue as to the age of deposition of the strata, merely representing the age when, after the last phase of metamorphism that affected the
rocks, the temperature decreased to a point at which radiogenic argon was
retained. They do, however, support the view that the formation of the schists
as we now see them was the result of a Caledonian metamorphic episode.

Western Series

The *Western Series* (Summ. Prog. 1932, p. 72) crops out in the north-west
corner of North Roe and consists of acid orthogneiss with sheets of basic and
ultrabasic gneiss. Pringle (1970) has subdivided this series into two groups, the
Uyea Group in the west and the *Wilgi Geos Group* in the east. The former is
composed of irregularly foliated gneiss made up mainly of orthoclase and
oligoclase with minor amounts of quartz, hornblende, biotite and epidote.
It contains some lenses of pyroxene-hornblende-granulite and is cut by many
veins of foliated pegmatite. Large sheets of metagabbro are well seen at
Fugla Ness and on the Isle of Uyea. The Wilgi Geos Group is separated from
the Uyea Group by the Uyea Shear Zone and consists of pale augen-gneiss
with thin sheets of fine-grained amphibolite. It differs from the gneiss of the
Uyea Group in that large parts of it have a strongly developed schistose fabric
which is parallel to the shear planes bounding the group. Locally the schistose
gneiss is brecciated or has mylonitic bands parallel to the schistosity.

Pringle (1970) believes that both these groups were originally granitic
complexes cut by gabbroic sheets and dykes. He also states that the complex
which now forms the Uyea Group was emplaced first and had perhaps
already undergone two periods of metamorphism before the Wilgi Geos
igneous complex was emplaced. Because both groups were affected by several
periods of metamorphism and deformation before the rocks of the two
eastern series were laid down they are believed to be of Lewisian age. The
Wilgi Geos Group was later involved in the Caledonian earth movements
which produced major shear zones and pushed individual thrust slices or
nappes westwards. The Uyea Group, on the other hand, appears to have
remained a stable block throughout that period. Pringle has suggested that
the Uyea Group may be part of the Lewisian foreland of the Caledonides and
that the Uyea and Wester Keolka shear zones (Fig. 4) may be branches of
the northward continuation of the Moine Thrust.

The *Ve Skerries* (north-west of Papa Stour on Fig. 2) consist largely of pale
grey to pink strongly banded albite-gneiss with varying proportions of mica
and hornblende. There are also some lenticular masses of foliated granite
and two thick bands of partially granitised hornblende-schist and mica-
schist. The Ve Skerries gneisses are unlike the metamorphic rocks of the
Walls Peninsula but are comparable to the orthogneisses forming the Wilgi
Geos Group of North Roe (Pringle 1970, p. 164). It is thus likely that the
Ve Skerries rocks form part of the Lewisian foreland, or, at least, a Lewisian
tectonic enclave within the Caledonian belt.

Fethaland Series and Ollaberry Series

To the east of the Wester Keolka Shear Zone lie two series of banded metamorphosed sedimentary, basic volcanic and pyroclastic rocks separated from

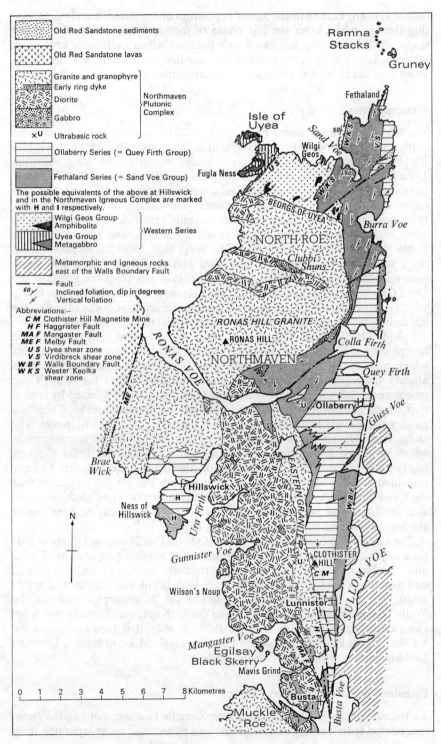

Legend:

- Old Red Sandstone sediments
- Old Red Sandstone lavas
- Granite and granophyre
- Early ring dyke
- Diorite — Northmaven Plutonic Complex
- Gabbro
- ×U Ultrabasic rock
- Ollaberry Series (= Quey Firth Group)
- Fethaland Series (= Sand Voe Group)

The possible equivalents of the above at Hillswick and in the Northmaven Igneous Complex are marked with **H** and **I** respectively.

- Wilgi Geos Group Amphibolite — Western Series
- Uyea Group Metagabbro
- Metamorphic and igneous rocks east of the Walls Boundary Fault
- Fault
- 60 Inclined foliation, dip in degrees
- Vertical foliation

Abbreviations:—
- **C M** Clothister Hill Magnetite Mine
- **H F** Haggrister Fault
- **MA F** Mangaster Fault
- **ME F** Melby Fault
- **U S** Uyea shear zone
- **V S** Virdibreck shear zone
- **W B F** Walls Boundary Fault
- **W K S** Wester Keolka shear zone

FIG. 4. *The metamorphic rocks and plutonic complexes of North Roe and Northmaven*

each other by a shear zone, named the Virdibreck Shear. These are, respectively, the Fethaland Series and the Ollaberry Series, whose distribution can be seen from Fig. 4. The major subdivisions of these groups, as recognised by Phemister (1976, *in* Summ. Prog. 1930), are set out in Table I. The succession is best seen in the northern part of North Roe where Pringle (1970, figs. 5 and 6) has established a more detailed lithological subdivision.

TABLE I. *The metasediments of north-west Mainland*

		Major Subdivisions	
		Phemister's Series	Groups in Pringle's Eastern Series
Components and Lithology			
East 6. Banded quartz- and muscovite-schist			
5. Calcareous Group ranging from quartzose schist to siliceous limestone		Ollaberry Series = Queyfirth Group	
4. Greenschist Group			
3. Banded quartz- and muscovite-schist			
2. Muscovite-schist with garnet and chloritoid in parts			
West 1. Banded hornblendic gneiss with granulitic gneiss and bands of impure quartzite.		Fethaland Series = Sand Voe Group	

The *Fethaland Series* is characterised by the presence of thick hornblendic gneisses, but also contains thick bands of impure quartzite, some of which are locally pebbly. Near its top there are thick lenses of amphibolite and metagabbro which probably originated as sills or flows of basic igneous rock. The series is coarsely crystalline and garnetiferous over most of its outcrop. The *Ollaberry Series* is generally finer-grained and only locally garnetiferous. Two of its groups have a distinctive lithology. These are the *Greenschist Group* which consists of fine-grained green-coloured schists with albite, epidote, chlorite and, locally, small garnets, and the *Calcareous Group* which contains calcareous schists with limestones and some bands rich in calc-silicate minerals. In the Lunnister area the Ollaberry Series contains a strongly deformed conglomerate and at Clothister Hill there is a bed of graphite-schist associated with a steeply inclined epigenetic deposit of magnetite enclosed in a skarn composed of garnet, hornblende, pyroxene and epidote. In Fethaland and Queyfirth the series contains intrusions of greenstone and serpentinite.

In the area around Colla Firth and Ollaberry the rocks of the Ollaberry Series are folded into a southward plunging synform. Farther south, at Lunnister, the structure of the rocks also seems to be a tight synform but, as the rocks become progressively more mylonitised and phyllonitised towards the south, their basic structural pattern becomes more difficult to unravel. Near the head of Busta Voe the belt of metamorphic rocks has decreased in width to a few hundred metres and it is here bounded and cut by a series of faults which splay out north-westwards from the Walls Boundary Fault.

Pringle has suggested that in North Roe the two series of metasediments were affected by two major phases of deformation. The first produced the Wester Keolka Shear Zone and was responsible for the pervading schistosity in the metasediments and the Wilgi Geos gneisses. It also produced some near-isoclinal intrafolial folds. The second deformation produced the other easterly dipping shear belts within the metasediments and gave rise to a series of folds. These folds are overturned to the west, suggesting that the rocks of the eastern series were being pushed westward over the 'basement.' Later dislocations include the development of open folds plunging to the north-west and south-east and the formation of the north-east trending faults. In the central and southern parts of the outcrop Phemister has recognised two major phases of deformation. The first is thought to have produced a large eastward-closing recumbent fold with shear zones along and parallel to its axial plane. The individual thrust slices appear to have been pushed to the west. The second phase was responsible for the rotation of this near-horizontal structure into a vertical position to give rise to the complex sheared synformal structure which we can recognise to-day. The brittle dislocations along the Walls Boundary Fault and its branches are ascribed to a later phase of large-scale, possibly transcurrent, movement (p. 69).

The main episode of prograde metamorphism affecting both the Fethaland and Ollaberry series probably coincided with the first phase of folding in these rocks and produced a mineral assemblage in the lower amphibolite facies. Subsequent mineral changes were due mainly to mechanical deformation which reduced biotite to chlorite, but these were so slight that pyroxene is still preserved in some of the mylonitic rocks. Exceptions to this pattern are found in a few small areas as in the skarn of the Clothister Hill magnetite where the presence of garnet and pyroxene indicates the local development of relatively high temperatures (p. 118). The effects of contact metamorphism by the Northmaven Complex are slight and very localised.

Metamorphic Rocks of the Ness of Hillswick

In the Hillswick area two series of metamorphic rocks are present. That occupying the southern part of the Ness of Hillswick consists largely of banded hornblende-gneiss and hornblende-schist with a band of silvery mica-schist and with sheared lenticles of serpentine and talc. This lithology is comparable to that of the Fethaland Series. The remainder of the outcrop contains a series of siliceous schists and granulites, muscovite-schists with garnet, kyanite and chloritoid and a band of hornblendic gneiss and granulite. Pringle has tentatively equated this series with his Queyfirth Group (i.e. Ollaberry Series) but because neither typical greenschists nor calcareous rocks are present this correlation is doubtful. The Hillswick rocks are strongly folded.

Metamorphic rocks within the Northmaven Plutonic Complex

There are three large and several smaller outcrops of metamorphic rocks partly or completely surrounded by the plutonic rocks which form most of the Busta Peninsula, the island of Muckle Roe and the adjacent smaller islands (Fig. 4). The metamorphic rocks form part of a banded group of

A. Lineated schist. Norby, Walls Peninsula, Shetland Mainland (D977)

Plate II

B. Ptygmatic folding. Aith Voe, Shetland Mainland *(D. Flinn)*

C. Porphyroblast schist. Gonfirth, Shetland Mainland *(D. Flinn)*

A. Muckle Flugga Lighthouse, Unst, with cliffs of Herma Ness behind *(Aerofilms Ltd.)*

Plate III

B. Funzie Conglomerate, south-east coast of Fetlar *(W. Mykura)*

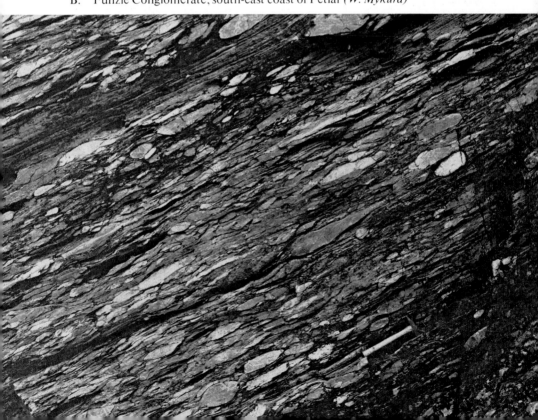

hornblende-rich schists in which pelitic mica-schists, locally garnetiferous, are important and quartzose granulites are minor constituents. Most of the rocks are similar to those found in the Fethaland Series. It is possible that some of the smaller masses of metamorphic rock are true inclusions or xenoliths within the igneous complex, but other masses, such as those at Egilsay and Black Skerry, are probably roof pendants. The metamorphic rocks of Muckle Roe and the Busta Peninsula could be either parts of the roof or floor of the adjoining igneous complex.

The rocks were regionally metamorphosed in the amphibolite facies and experienced subsequent shearing and fracturing at low temperature prior to being intruded by the igneous complex. Later thermal metamorphism adjacent to basic rocks of the complex locally reached pyroxene-hornfels facies, and andalusite and cordierite were extensively developed in the pelites. The thermal alteration was followed by low-temperature hydrothermal changes which led to the alteration of cordierite and andalusite to pinite. Still later the rocks were subjected to local potassium metasomatism associated with the intrusion of the granite, and finally the masses close to the Walls Boundary Fault were both shattered and intensively scapolitised.

Metamorphic rocks of the Walls Peninsula and Foula

Metamorphic rocks form the northern coastal strip of the Walls Peninsula and parts of the islands of Vementry (Fig. 5) and Papa Little. In these the following four lithological units, listed from north to south, have been recognised:

1. *Vementry Group.* Hornblende-schist and amphibolite with bands of semi-pelite and some quartz-granulite.
2. *Neeans Group.* Platy feldspathic muscovite-schist with lenticular masses of coarse hornblende-schist and some thin bands of limestone and epidote-clinozoisite rock.
3. *West Burra Firth Group.* Tremolite- and phlogopite-schist with calc-schist and limestone.
4. *Snarra Ness Group.* Mainly hornblende- and mica-schist with bands and lenses of amphibolite.

The outcrops of these groups do not form continuous bands and some of the boundaries between the groups appear to be shear planes. A high proportion of these rocks were originally siltstones, shales and sandstones. Calcareous mudstones and thick limestones probably formed an appreciable part of the West Burra Firth Group, while basic lavas and/or pyroclastic deposits were abundant in the two hornblendic groups. The thick masses of hornblende-schist and amphibolite may have been basic sheets or sills. The rocks now contain many veins of granite and pegmatite as well as lits and porphyroblasts of feldspar. The granitic material is particularly abundant in the hornblendic rocks and forms up to 20 per cent of the total volume of the Snarra Ness Group.

The tectonic history of this belt of metasediments is similar to that of the North Mainland belt. During the earliest recognisable phase of deformation the banding of the rock was folded into a series of tight minor folds with axial planes parallel to the regional foliation. This was the main period of folding

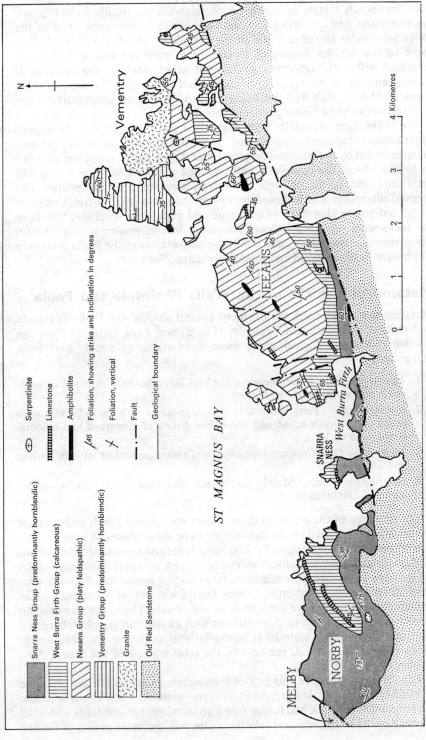

FIG. 5. *The metamorphic rocks of the Walls Peninsula*

and it produced a pronounced lineation parallel to the axes of the folds. This phase was followed by a local, relatively minor, episode of folding during which a weak axial-plane strain-slip cleavage was formed in some parts of the Snarra Ness Group. Meanwhile, or perhaps somewhat later, the Vementry Group, which now forms the northern part of the outcrop, was subjected to a phase of intense shearing with some mylonitisation, indicating that a major shear zone was probably being formed just north of the present outcrop. Later dislocations include the development of belts of conjugate kink-folds and more widespread kink-bands, all of which have an east-south-easterly axial trend. As the overall dip of the foliation of the metamorphic rocks is roughly parallel to the bedding of the overlying Old Red Sandstone strata, it follows that the foliation in the former must have been more or less horizontal when these sediments were laid down. The present southerly dip of the foliation resulted from the late Middle or Upper Devonian earth movements, described on p. 61.

The peak of metamorphism appears to have been attained during or just after the main period of folding. Over large parts of the area this took place under conditions transitional between the greenschist and amphibolite facies. Common minerals developed at this stage include garnet, which is abundant throughout the series, oligoclase, hornblende, tremolite, phlogopite, epidote and clinozoisite. There are a number of areas where diopside occurs in lime-stones and in calc-silicate rocks, indicating that, locally, higher temperatures may have prevailed. These thermal 'highs' occur in areas of intense granite veining and may be associated with the emplacement of these veins. Retro-grade metamorphism occurred both during the period of shearing in the Vementry Group and, to a lesser extent, during the phase of conjugate folding. The shearing produced extensive granulitisation and local mylonitisation of the rock-fabric and led to the breakdown of garnet and biotite to chlorite.

Foula. The metamorphic rocks forming the fault-bounded strip along the east coast of the island of Foula (Fig. 12) are made up of bands of garneti-ferous psammitic granulite and garnetiferous mica-schist. They contain lenses of amphibolite and epidote-rock as well as two thin bands of crystalline lime-stone. All the metamorphic rocks are cut by granite veins and contain lits and porphyroblasts of feldspar. There is a marked similarity both in the lithology and in the structural and metamorphic history of these rocks and the Neeans Group of the Walls Peninsula.

REFERENCES

Miller and Flinn 1966; Mykura and Phemister 1976; Phemister 1976; Pringle 1970; Walker 1932.

C

3. METAMORPHIC ROCKS OF SHETLAND:
II EAST MAINLAND AND ADJACENT ISLANDS

East Mainland Succession

Most of the metamorphic rocks of Shetland Mainland east of the Walls Boundary Fault belong to the East Mainland Succession (Plate IV), a more or less continuous sequence of steeply inclined to vertical north to north-east trending metasediments, which may be 22 to 27 km thick. The outcrop of this group of rocks is cut by the Nesting Fault, a major transcurrent fault which may have a dextral displacement of 16 km. The succession can be divided into four main divisions, the thicknesses and lithological units of which are shown in Fig. 6. As the thicknesses are based only on the width of the outcrops and the dip of the beds, and do not allow for the many small-scale structural complexities, they must be regarded as approximate. In spite of the paucity of way-up criteria within the beds it is believed that the oldest beds lie in the west and that the succession gets progressively younger in an eastward direction.

Yell Sound Division

The Yell Sound Division, which is truncated in the west by the Walls Boundary Fault, crops out along the shores of Sullom Voe and forms a narrow strip along the east coast of Aith Voe. It is also found on the east shore of Swining Voe and probably extends across Yell Sound to include the rocks of Yell (p. 32). It is composed almost entirely of highly feldspathic psammites (originally arkoses or feldspathic sandstones), which have been widely but very variably migmatised. There are two or three quartzite horizons within the sequence. The junction with the Scatsta Division to the east is not clearly marked because of interbanding and the masking effects of migmatisation.

Scatsta Division

The Scatsta Division can be divided into a lower *pelitic group* and an upper *quartzitic group* (Fig. 6). The former consists mainly of pelitic schist some of which contains large crystals of staurolite. The quartzitic group, which forms the prominent topographic ridge in central Mainland just west of Weisdale Voe and the Kergord valley, consists of pure and impure quartzites interbanded with medium- to fine-grained mica-schists. Individual bands in this group vary in thickness from a centimetre to several metres. They are highly deformed and metamorphosed and do not show any original sedimentary structures other than compositional banding. The quartzites are generally predominant but certain pelitic bands containing large staurolites can be traced for long distances along the strike.

The Yell Sound and Scatsta divisions contain a north to north-east trending, somewhat irregular belt of migmatite which has been termed the *Scatsta Permeation Belt* by Flinn (1954, p. 187). Within this belt the mica-schists have been converted into schistose gneisses, composed of small streaks and eyes of quartz and feldspar separated by anastomosing micaceous folia. The belt is bounded by two very persistent bands, on the west by augen-gneiss and on the east by schist with large porphyroblasts of microcline. The pelitic rocks associated with the belt contain such minerals as andalusite, shimmer aggregate, sillimanite and secondary muscovite, which indicate that the formation of the gneiss was accompanied by the thermal metamorphism of the adjoining rocks.

Whiteness Division

The Whiteness Division consists mainly of flaggy micaceous quartz-feldspar-granulite (psammite) and contains four thick and several thinner bands of crystalline limestone. The psammite is characterised by the presence of micaceous laminae which are regularly spaced at intervals of a few milli-metres. These laminations appear to be original features of the sedimentary rocks. *The Weisdale Limestone*, the basal member of the Division, is over 400 m thick and gives rise to a prominent topographic depression which can be followed from Weisdale northwards to Dales Voe. The *Whiteness Lime-stone*, which is up to 600 m thick in White Ness, thins out northwards. The *Girlsta Limestone*, which is persistent in the north and forms a small ridge in the Tingwall Valley, appears to lens out southwards on Trondra. All these lime-stones are calcite marbles which contain small quantities of epidote, zoisite, white mica and pyrite concentrated in fairly continuous layers. Epidote-rich lenses and thin semipelitic bands are present throughout the psammites of this division, but highly micaceous rocks are largely confined to one group, known as the *East Burra Pelite*.

The thick sequence of rocks, including the East Burra Pelite, between the Whiteness and Girlsta limestones contains a belt of migmatitic gneiss which has been termed the *Colla Firth Permeation Belt* by Flinn (1954). This is up to 2·5 km wide and can be traced from Swining Voe in Delting southwards to South Havra, a distance of 43 km. Its probable extension on the east side of the Nesting Fault passes through Nesting and extends from Cat Firth to Stava Ness. The gneiss contains much material of metasedimentary aspect and in most places it has retained its banded structure. The process of conversion to coarse-grained gneiss has acted very unevenly on the various rock types; psammitic and semipelitic rocks are coarser-grained than normal and contain abundant quartzo-feldspathic bands and lenses; the pelitic rocks have conspicuous micro-augen (small 'eyes') of plagioclase which are arranged with their long axes parallel to the schistosity, but calcareous and hornblendic rocks are preserved intact. As in the Scatsta belt the development of the gneiss is associated with the production of high-temperature minerals such as sillimanite.

Deformed 'veins' of granite, pegmatite and quartz-tourmaline rock are very abundant within and just outside the migmatite belt. Most of the granite

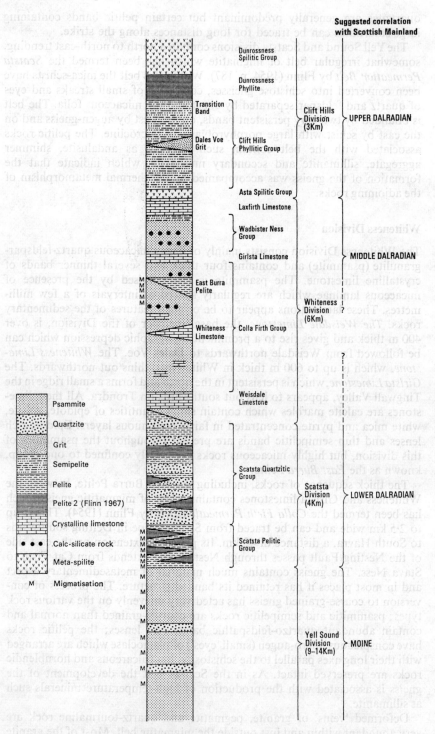

FIG. 6. *The East Mainland Succession—(graphic representation of stratigraphic units)*

'veins' are concordant sheets and lenses with sharp margins against the gneiss, but cross-cutting veins are also common. The largest granite mass has a maximum thickness of 500 m and extends for 6·5 km along the strike. Many of the concordant sheets are boudinaged and most have a marked schistosity which is parallel to, and continuous with, the regional foliation. The cross-cutting veins have been folded by buckling.

The two most easterly members of the Whiteness Division, the *Wadbister Ness Group* and the *Laxfirth Limestone*, are best seen in the area east of the Nesting Fault. The former is also well exposed around the head of Laxo Voe in Delting, just west of the fault, and forms the peninsula of South Nesting as well as Wadbister Ness. It consists of alternating bands of flaggy micaceous psammite and semipelite, frequent ribs and striped bands of calc-silicate rock and impure marble, and some bands of hornblende-schist. The *Laxfirth Limestone* may be more than half a kilometre thick. It is lithologically similar to the other limestones, but finer-grained.

Clift Hills Division

The Clift Hills Division follows conformably to the east of the Laxfirth Limestone. Its predominant rock types are chlorite- and biotite-muscovite-phyllites, but there are several important bands of spilite and some thick beds of grit. Along the western side (? base) of the division the *Asta Spilitic Group* is developed along the ridge east and north-east of Scalloway and close to the east shore of Lax Firth, and can be recognised further north on the islands of Gletness and on Hoo Stack. It consists of metasedimentary phyllite together with hornblendic phyllites and coarse-grained hornblendic rocks, which may have originated respectively as basic lavas and/or pyroclastic deposits and as basic intrusions.

The *Clift Hills Phyllitic Group*, though on the whole very uniform, contains rocks which range from pelites to impure quartzites. The phyllites are laminated on a millimetre scale, the darker, more micaceous layers representing metamorphosed mud, while the lighter are probably derived from sand and silt. It is occasionally possible to detect small-scale cross-bedding in these rocks. All the phyllites contain muscovite together with either reddish brown biotite and/or one of two varieties of chlorite. Some are graphitic. Within the group there are several thick units of coarse feldspathic grit, the most important being the *Dales Voe Grit* which can be traced from Dales Voe southwards to the latitude of Quarff.

The Clift Hills Phyllitic Group passes eastward through a transition bed into the *Dunrossness Phyllitic Group*. The latter can be followed from the latitude of Scalloway southwards along the eastern slopes of the Clift Hills to Scousburgh, where its outcrop turns sharply westward and then northward. The group also forms a large part of the Fitful Head peninsula. These phyllites are very uniform in texture and composition and they are characterised by abundant chlorite and muscovite and the almost invariable presence of chloritoid. Kyanite occurs in quartz segregations. In the metamorphic aureoles (p. 45) they also contain staurolite. *The Dunrossness Spilitic Group*, which crops out at Cunningsburgh, Channerwick and on the western cliffs of Fitful Head, contains metamorphosed lavas and pyroclastic rocks in which the original textures are locally well preserved. Pillow structure has been noted but is not common. At Cunningsburgh there is a considerable mass of serpentinite extensively altered to talc-magnesite-schist and intimately associated with the lavas and pyroclastics. The spilites are intercalated with metasediments, mainly graphitic phyllites, gritty impure quartzites and re-crystallised cherts. The spilites and cherts of this group are thought by Garson and Plant (1973) to be part of an ophiolite complex and to mark the position of a subduction zone.

Other Rock Groups

There are a number of areas in south Mainland, bounded either by tectonic dislocations or by the sea, which consist of rocks that resemble certain members of the East Mainland Succession but have not been definitely correlated with them (Flinn 1967a). The *Bigton Grits*, for instance, are a series of well-bedded quartzitic grits with lenses of hornblende-feldspar rock. They crop out along the western coastal strip between Maywick and Scousburgh and resemble the Dales Voe Grit (p. 27). It is likely that they are part of a fault-bounded slice of a grit in the Clift Hills Phyllitic Group. The *Garths Ness Hornblendic Series* forms the peninsula of Garths Ness, just east of Fitful Head. It consists of a shear-bounded mass of striped epidotic hornblende-schist with isolated beds of semipelite. It is at least 500 m thick, far thicker than any other hornblende-schist on Shetland Mainland.

Quarff Nappe and the Mélange

Between Rova Head, 4 km N of Lerwick, and Cunningsburgh, the East Mainland Succession is bounded on the east by a poorly exposed zone of tectonic dislocation which is probably made up of many shear-bounded slices of rocks similar to those found in the relatively unsheared ground both to the east and west. There are slices of graphitic phyllite, quartzite, staurolite-schist, altered serpentinite, marble and grit. This zone is believed to be a 'mélange' or schuppen-zone which was formed when a major nappe, now represented by the *Quarff Succession* (Plate IV) overrode the rocks of the East Mainland Succession. The Quarff Succession is best exposed around Quarff and Fladdabister where four main rock groups can be recognised. Three of these can be fairly confidently correlated with rock types in the East Mainland Succession. Thus the most westerly belt of rocks in the Quarff area consists of gneisses which are similar to the gneisses in the Colla Firth Permeation Belt. To the east of these a thick series of flaggy psammites, semipelites and calc-silicate rocks is closely similar to the schists of the Wadbister Ness Group, and a limestone, which crops out south-west of Fladdabister, has been tentatively correlated by Flinn with the Laxfirth Limestone. East of this limestone there is a thick series of bedded grits called the *Fladdabister Grits*. These are different from any of the rocks in the East Mainland Succession, and they consist of quartz and acid plagioclase grains set in a matrix which ranges from calcareous to siliceous. Flinn has suggested that these grits may never have been part of the East Mainland Succession, and that they are part of a rock slice which might be equated with a slice of similar grit just east of the mélange at Rova Head.

Structure

Only one period of intense deformation, termed the *Main Deformation* (Flinn 1967a), is recognised in the East Mainland Succession. This was responsible for the formation of the tectonic fabric in the rock which ranges from predominantly planar (s-tectonite) to linear (l-tectonite). During this phase small tight and isoclinal folds of the bedding were formed in many parts of the succession, but major folds have not been recognised with the possible exception of a big fold picked out by the outcrop of the Dunrossness Phyllite in the Channerwick–Scousburgh–Maywick area. The foliation of the rock is determined by the limbs of the isoclinal folds; a schistosity lying parallel to this foliation is present in the more micaceous beds. In the migmatite belts the foliation is less regular and is commonly distorted around resistant lenticular masses. Many hornblende-schist bands are boudinaged and most rocks display a prominent rodding or mineral lineation.

Structures associated with later phases of deformation include minute crinkles in some micaceous beds, attributed to weak tectonic movements, and small kink-bands and conjugate folds, locally with broken axial planes and shattered short limbs. These were formed by a late phase of folding which took place under brittle conditions. There is also a very small number of large open folds which may be attributed to the forceful emplacement of the Spiggie and Channerwick granites (p. 45).

At a later date the East Mainland Succession was cut by a number of subparallel faults of which the Walls Boundary Fault and the Nesting Fault, both probably with large dextral displacements (p. 69), are the most important. The Quarff Nappe was emplaced before the deposition of the East Shetland Old Red Sandstone and probably before the main movement along the faults associated with the Nesting Fault.

Metamorphic History

The metamorphic and structural history of parts of the East Mainland Succession has been studied by Flinn and May. Flinn (1954) suggests that in Delting (north Mainland) the earliest recognisable episode after the deposition of the sedimentary rocks was a period of regional metamorphism. This coincided with the main phase of deformation which imprinted the dominant planar and linear tectonite fabric on the rock. He believes that at that time the grade of metamorphism increased westwards and that most of the Scatsta Division lay within the kyanite zone, whereas the Whiteness Division, which contains much less garnet, was low in the garnet zone or even below. The next episode was the development of the Scatsta and Colla Firth permeation belts. This took place after the period of deformation and its associated regional metamorphism, but while thermal gradients generated within the rocks at that time were still in existence. The location of the belts of permeation was in no way controlled by the regional metamorphism, as they cut across the pre-existing isograds and follow the structural and lithological pattern of the rocks. The permeation was accompanied by thermal metamorphism in the immediate neighbourhood of the belts. Flinn concludes that in Delting the regional metamorphism and permeation were two largely independent processes.

Flinn (1967a) recognises a very similar sequence of events in the southern part of Shetland Mainland, where he notes the following metamorphic episodes:

1. *Tectonising metamorphism* (i.e. metamorphism associated with the tectonic phase). This coincided in time with the principal period of deformation. During this period the platy and elongate minerals developed a preferred orientation which determined the foliation and lineation of the rocks. Minerals formed at this stage are biotite, chlorite, muscovite, quartz, feldspar and possibly garnet.

2. *Porphyroblast metamorphism*. During this phase a number of minerals, normally developed as a result of regional metamorphism, crystallised as porphyroblasts. The orientation of these minerals is random and quite unrelated to the tectonite fabric of the rocks, thus indicating that their growth was later than the 'tectonising metamorphism'. The new minerals formed were biotite, chlorite, staurolite, kyanite, chloritoid and garnet.

3. *Permeation metamorphism*. This is the episode when the Colla Firth Permeation Belt was formed and the associated granite and pegmatite veins were emplaced. It took place while the thermal gradients of the regional metamorphism (phases 1 and 2) were still in existence. The permeation is believed to have taken place under static conditions, the alignment of many minerals parallel to the regional fabric being ascribed to mimetic crystallisation. As in Delting, the permeation produced a phase of localised thermal metamorphism, during which diopside and microcline were developed in the calc-silicate bands. The mineral associations in these thermally altered zones indicate conditions characteristic of the boundary petween the pyroxene hornfels and hornblende hornfels facies.

4. *Thermal metamorphism*. The emplacement of the Spiggie, Channerwick and Cunningsburgh granites (p. 45) produced extensive thermal aureoles within which such minerals as staurolite, chloritoid, andalusite, kyanite, sillimanite, garnet and muscovite were developed in the pelitic rocks of the Dunrossness Phyllitic Group.

A somewhat different sequence of metamorphic episodes was deduced by May in his study of the area around Scalloway. Here the textural evidence suggests that prograde metamorphism took place in three stages. During the first stage kyanite and staurolite were formed in pelites, but most traces of this stage have been obliterated by later events. Most minerals now present in the rocks were formed during the second metamorphism which coincided in time with the main phase of deformation. This was also the period when the Colla Firth migmatite belt was formed and the numerous associated cross-cutting and concordant granite and pegmatite sheets and veins were emplaced. The crucial evidence for the belief that the migmatite emplacement took place while deformation was still in progress comes from the granites and pegmatites. In these rocks it can be shown that the constituent minerals have been granulated and recrystallised to produce a tectonite fabric which is continuous with the fabric of the country rock. The only evidence for a period of post-tectonic metamorphism is provided by the presence of diopside in calc-silicate rocks, some of which crystallised under apparently static conditions.

Lunnasting, Whalsay and Out Skerries

The metamorphic rocks forming the peninsulas of Lunna Ness and Lunnasting and the island of Whalsay consist of highly metamorphosed homogeneous pelitic and semipelitic gneisses with much quartzite and some impure lime-stones. Their inclination ranges from vertical in the west to horizontal in Whalsay in the east and their trend is generally north-north-easterly. Nearly all these rocks have been intensely migmatised and are now coarsely crystal-line. In Lunna Ness some are almost completely granitised, showing only vague relicts of their original structure, and a continuous band of very coarse microcline augen-gneiss extends along the centre of this peninsula. In the Lunning district the gneiss has many prominent crystals of staurolite and kyanite, now mostly altered to shimmer aggregate. The most common minerals forming the gneiss are feldspar, biotite, muscovite, quartz and garnet. A thick band of limestone with calc-silicate ribs crosses Lunnasting from Dury Voe to the head of Vidlin Voe and there are a number of lenses of hornblendic gneiss, particularly in Lunna Ness. Whalsay is formed of pelitic gneiss with abundant garnet and kyanite, and some quartzites, cut by numerous small granite intrusions. A belt of very intense migmatisation extending along the south-east coast of the island has produced granite-like rocks.

The Out Skerries consist of a variety of rock types which have a trend that swings from north-east in the extreme west to east-north-east in the centre and east of the island group. The rocks also have a strong linear fabric due to intense folding, on all observable scales, on axes that are nearly horizontal. A thick crystalline limestone passes through the centre of the island group. This is bounded on either side by semipelitic granulites, calc-silicates and some pelites which are usually partially migmatised and commonly contain sillimanite. The migmatisation is most extensive along the southern shores of the island group. All the rocks have been invaded by pegmatite and granite veins and ribs, some of which are several hundred metres long. The edge of a somewhat larger granite intrusion occurs in the extreme north of the group.

No published modern work on the structure, metamorphic history and possible stratigraphic correlation of the Out Skerries is available. Robertson (*in* Summ. Prog. 1934, p. 71) thought that these rocks bear a strong resemblance to those forming the eastern margin of the main permeation belt in South Nesting. If this correlation is accepted, the Out Skerries rocks form the north-eastward continuation of part of the Whiteness Division. Flinn believes that the rocks on both sides of the thick limestone are similar to the Wadbister Ness Group, but states that the migmatisation, which is of an unusual type, is less strongly developed than in the Colla Firth Permeation Belt. The Skerries limestone may thus be the Laxfirth Limestone which here occupies the core of a tight fold and is flanked by rocks of the Wadbister Ness Group.

The islands south of the main group of Out Skerries contain rocks closely resembling the Asta Spilites of Shetland Mainland, associated with gritty quartzites with flute casts and with phyllites which contain some staurolite. All these rock types are closely similar to the rocks of the Clift Hills Division.

REFERENCES

Flinn 1954, 1967a; Flinn and others 1972; May 1970; Miller and Flinn 1966.

4. METAMORPHIC ROCKS OF SHETLAND: III YELL, UNST AND FETLAR

The northern isles of Shetland contain two major geological units. The western unit (Fig. 2) comprises the whole of Yell, the western half of Unst and the most westerly peninsula of Fetlar. It consists of highly metamorphosed and, to a large extent, intensely migmatised metasedimentary gneisses which have some characteristics in common with the migmatised lower divisions of the East Mainland Succession. The eastern unit comprises the eastern half of Unst and most of Fetlar and is separated from the rocks to the west by a major thrust zone. It consists of a number of shear-bounded blocks of serpentinite, metagabbro and phyllite which are part of two major nappes that were emplaced at a high structural level. The tectonised phyllites between and below the nappes consist of slices or 'schuppen' of rock cut from the basement and from the nappes, and of newly-formed sediments derived by erosion from the nappes and the basement. These rocks were converted to their present phyllitic state at the time of nappe emplacement by retrogressive metamorphism of the slices of gneiss, hornblende-schist and metagabbro and by prograde metamorphism of the newly deposited sediments. The western metasediments of the northern isles are collectively termed the *Basement* and the eastern nappes are called the *Nappe Pile* (Flinn 1958).

Basement (and Saxa Vord Block of Unst)

Yell

The greater part of Yell consists of garnetiferous mica-plagioclase-gneiss which has been tentatively correlated with the Yell Sound Division of Shetland Mainland. Its foliation trends north-north-west in the north of the island, roughly north–south in the centre and north-north-east in the south. The dip of the foliation is generally steep and its overall inclination is to the west, but in the south-east corner of the island it is vertical or steeply inclined to the east.

The north-eastern coastal strip of Yell, the island of Hascosay (Fig. 7) and the south-eastern corner of Yell are underlain by striped granulitic oligoclase-gneiss which contains large masses of hornblende-gneiss, some bands of calc-silicate rock and a band of gneiss with augen of microcline. These outcrops form part of a strip of rock which seems to be separated from the main mass of Yell by a major dislocation which could be a branch of the Nesting Fault. It may well form part of the structural unit which includes the western (foreland) blocks of Unst and Fetlar (pp. 33–36).

West of the eastern coastal strip of Yell a sequence of six poorly defined lithological units can be recognised, but none of these form continuous belts

32

throughout the length of the island. The most easterly belt exposed on the east coast between Basta Voe [53 95] and Mid Yell Voe [52 91] consists of mica-schist with small lenticular masses of siliceous granulite. This is followed westwards by a more massive gneissose rock with many veins of pegmatite, which is in turn succeeded by flaggy siliceous and semipelitic schists which form the Hills of Lussetter [52 90] and Vatsie [52 88] south of Mid Yell Voe. The centre of the island is traversed by silvery garnetiferous mica-schists (semipelites) which contain a thick lenticular quartzite. The latter gives rise to the highest ground in Yell, which includes the Hill of Arisdale [50 85] and the Ward of Reafirth [51 88]. West of this horizon there is a wide belt of garnetiferous oligoclase-mica-gneiss (semipelite) which is in places intensely migmatised and contains many coarse red granite and oligoclase-pegmatite veins. This migmatised belt takes in the northern and north-western parts of Yell and contains some areas in which sillimanite and tourmaline are developed. The western coastal cliffs of Yell consist mainly of evenly foliated muscovite-biotite-gneiss together with some silvery mica-schists, but the stretch between West Sand Wick [44 89] and the Ness of Sound [45 82] contains, in addition, lenses of hornblendic and siliceous gneiss and very abundant dykes of pegmatite.

Unst

The western part of Unst consists of two principal tectonic units or 'blocks' known as the Valla Field and Saxa Vord blocks (Read 1934c). These are separated from each other and from the Nappe Pile to the east by major shear zones full of lenticular rock-slices. The Saxa Vord Block is not strictly part of the Basement (see p. 35) and is included under this heading only for ease of description. The principal lithological units of these blocks are set out on p. 34[1] (see also Fig. 7a).

The *Valla Field Block* forms the hilly western part of Unst and is bounded on the east by a belt of dislocation, which is now followed by the prominent valley, marked by a string of lochs, that extends southwards from the head of Burra Firth to Belmont. The foliation of the rocks in this block has an overall north-north-easterly trend and is generally inclined at varying angles to the east-south-east. Most of the outcrop is thought to form part of the eastern limb of a major north–south trending antiform, the axial trace of which passes through the outcrop of the Westing Group. The rocks forming the block are mostly of sedimentary origin, but at Lunda Wick [56 04] there are some zoned ultrabasic bodies which grade outwards from a core of antigorite to actinolite, chlorite and biotite. The rocks have been extensively migmatised and intruded by a network of sills and sheets of granite and pegmatite and later by dykes and sills of spessartite.

Read (1934c, 1937) has shown that the block was affected by three episodes of metamorphism. The first was a period of high-temperature, high-stress prograde regional metamorphism which affected the whole block, and was

---

[1] This table is based on Read's (1934c) interpretation of the structure. Flinn (1958) suggests that groups 1, 4 and 5 of the Saxa Vord Block are not part of this block, but belong to the Lower Schuppen Zone of the Nappe Pile (see p. 38 and Fig. 7b).

Valla Field Block

East: 3. *Burra Firth Group:*   Pelitic, siliceous and feldspathic gneiss, with belts of garnet-staurolite-tourmaline-gneiss and rare calc-silicate rocks.

         2. *Valla Field Group:*   Pelitic gneiss, passing eastward from coarse biotite-permeation-gneiss with lenses of hornblende-schist in the west into staurolite-kyanite-garnet-gneiss with thin siliceous and semipelitic bands.

West: 1. *Westing Group:*   Pelitic, calcareous and hornblendic gneisses, containing the thick *Westing Limestone*, a crystalline limestone rich in calc-silicate minerals.

Saxa Vord Block

East   5. *Norwick graphitic schists*

         4. *Norwick hornblendic schists*

         3. *Saxa Vord Schists:*   Pelitic schist with chloritoid, kyanite, staurolite and garnet, some bands of quartzo-feldspathic rock.

         2. *Queyhouse Flags:*   Siliceous and quartzo-feldspathic flags with subordinate mica-schists.

West: 1. *Loch of Clift Limestone*

connected with the main epoch of deformation. At this stage kyanite, staurolite, garnet and biotite were formed in the pelites; and diopside, phlogopite, epidote, idocrase and plagioclase in the calcareous rocks. The widespread migmatisation and veining by pegmatite and granite may have accompanied or closely followed this phase. The second episode was a retrograde metamorphism which affected only the southern half of the block. At this time the minerals staurolite and garnet were partially or completely converted to chloritoid and chlorite, and in the calcareous rocks diopside was altered to tremolite. The intensity of retrograde metamorphism increased towards the south. The third episode was a dislocation metamorphism which was confined to the immediate vicinity of the shear zone bounding the block in the east. At this stage the minerals formed during the earlier metamorphic episodes were broken down into chlorite and a chlorite-phyllite was produced.

The *Saxa Vord Block* forms the high hills of northern Unst just east of Burra Firth, but thins southward from there to form (according to Read) a narrow irregular belt between two shear zones which converge just south of the Loch of Watlee [60 05]. In the northern part of the outcrop the main rock types are dark grey chloritoid-kyanite-schists (the Saxa Vord Schists) and quartzo-feldspathic flags (the Queyhouse Flags). The schists must be of considerable thickness and their foliation dips at a moderate angle to the east-south-east. Further south the various lithological units are close together and each unit may be a separate tectonic lens (see p. 38). The rocks of the Saxa Vord block are not migmatised and pegmatite veins are confined to the Queyhouse Flags. The metamorphic history of the block, based largely on evidence from the

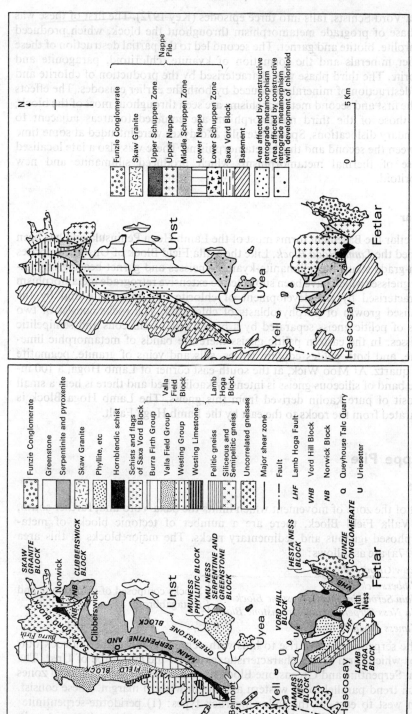

FIG. 7. (a) *Geological sketch-map of Unst and Fetlar*
(b) *Structural map of Unst and Fetlar*

Saxa Vord Schists, falls into three episodes (Key 1972). The first of these was a phase of prograde metamorphism throughout the block, which produced staurolite, biotite and garnet. The second led to the partial destruction of these earlier minerals and the formation of kyanite, chloritoid, paragonite and chlorite. The third phase was characterised by the production of chlorite and the destruction of minerals produced by both the earlier episodes. The effects of the first and second metamorphisms are seen throughout most of the block, but those of the third metamorphism are confined to areas adjacent to boundary dislocations. Spessartite sills and dykes were intruded at some time between the second and third metamorphisms. There was also a late localised phase of thermal metamorphism, which produced sillimanite and new chloritoid.

Fetlar

In Fetlar the Basement forms most of the Lamb Hoga Peninsula and has been termed the *Lamb Hoga Block*. Like the Valla Field Block of Unst it comprises high-grade staurolite-sillimanite-kyanite-gneisses and garnet-bearing migmatitic gneisses which have been subjected to extensive retrograde metamorphism characterised by the development of chlorite and muscovite and by the localised growth of porphyroblasts of chloritoid. The rocks consist of two belts of pelitic gneiss separated by a broad belt of siliceous and semipelitic gneisses. In the eastern pelitic gneiss there are bands of metamorphic limestone, and both pelites contain lenses, sills and veins of granite, pegmatite and quartz. At Moo Wick, at the south-east corner of Lamb Hoga, a 100 m-wide band of siliceous gneiss is intensely kaolinitised and there is here a small deposit of pure kaolin derived from this gneiss. The Lamb Hoga Block is separated from the rocks to the east by the Lamb Hoga Fault.

Nappe Pile

Unst

East of the zone of movement which limits the Saxa Vord and, further south, the Valla Field Block, there are a number of tectonic blocks of metamorphosed igneous and sedimentary rocks. The major blocks in this area (Fig. 7a) are as follows:

Skaw Granite Block

Clibberswick Block
Main Serpentine and Greenstone Block } Blocks composed of serpentinite and
Mu Ness Serpentine and Greenstone Block } metagabbro

Muness Phyllitic Block

The serpentinites give rise to hills with ochre-brown-weathering rock outcrops which form such a characteristic feature of the landscape of Unst. The Main Serpentine and Greenstone Block has a number of compositional zones which trend parallel to its western and north-western margin. These consist, from west to east, of the following rock types: (1) peridotite-serpentinite; (2) dunite-serpentinite; (3) pyroxenite, and (4) metagabbro ('greenstone'). This block appears to be part of a tilted weakly metamorphosed igneous mass whose lower (western) ultrabasic part may have been differentiated *in situ*, but whose upper, basic, part was possibly intruded slightly later. The serpentinite contains pockets of chromite which were formerly quarried (p. 118). The ultrabasic rocks are altered along the thrusts bounding the blocks, and along certain dislocations within them, into antigorite-serpentinite, talc and, locally, chlorite-schist. Talc of commercial quality occurs at Queyhouse just east of Loch of Clift and also along the margin of the Clibberswick Block at Clibberswick (p. 119).

The Skaw Granite Block is bounded on the west by a major shear zone. It consists of a pink foliated augen-granite, the augen being red potash-feldspars which reach a length of 8 cm and are set in a granulitic base of quartz and mica. Over most of its outcrop the augen-granite contains large and small xenoliths of metasediment, mostly siliceous granulites and ghosts of pelite. According to Read (1934c, p. 677) the foliation of the granite, the orientation of the feldspar phenocrysts and the attitude of the stratification of the xenoliths are all coincident, and have an overall north-easterly to easterly strike. Some of the xenoliths contain feldspar porphyroblasts which are similar to the large feldspars in the granite.

The Muness Phyllitic Block forms a large outcrop of phyllite with rare bands of schistose conglomerate. The rocks within it are of low metamorphic grade, and they have not been subjected to an earlier high-grade metamorphism. The phyllites have a typical 'woody' fibrous structure which is due to intense microfolding combined with a linear fabric. Later kink-folding of the lineation is common. In addition to the blocks enumerated above there are two small tectonic blocks composed of hornblendic and graphitic schists and phyllites in the Norwick area. The individual blocks of Unst are in places separated by very fine-grained silvery phyllonites, which are parts of the schuppen-zones and are essentially mylonitised pelitic gneisses.

Fetlar

The greater part of Fetlar consists of a number of shear-bounded thrust blocks of serpentinite, metagabbro and phyllite which bear a complex structural relationship to each other (Fig. 7a). There are three serpentinite blocks termed from west to east the *Hamars Ness Block* (serpentinite and metagabbro), the *Vord Hill Block* (peridotite-serpentinite) and the *Hesta Ness Block*. The serpentinites are generally ochreous-weathering but contain varying amounts of bastite which produces a rough-weathering rock. As in Unst the serpentinite is altered to antigorite-serpentinite and locally to talc close to the bounding shear planes. The small Hesta Ness Block in north-east Fetlar is completely altered to antigorite and steatite and contains bands of pure talc which have been exploited commercially in the past. In addition to the serpentinite blocks there are a number of highly complex shear-bounded zones which consist of varying proportions of phyllite, strongly lineated graphite-schist, phyllitised metagabbro ('greenstone'-phyllite) and schistose conglomerate with small, highly elongated pebbles. These zones also contain small masses of serpentinite, metagabbro and migmatitic gneiss. Of particular interest are the schistose conglomerates which are best seen along the coast at Uriesetter west of Vord Hill and which contain cigar-shaped pebbles of metagabbro (indistinguishable from that in the Hamars Ness Block), quartz-albite-porphyry, granophyre and material derived from the underlying sediment.

Most of the eastern peninsula of Fetlar is occupied by a great thickness (at least 1200 m) of conglomerate. This conglomerate, known as the *Funzie Conglomerate*, consists of strongly deformed pebbles of quartzite with subordinate biotite-albite-granite, granophyre, quartz-albite-porphyry and, locally, basic igneous rock, and rare pebbles of marble. These are set in a phyllitic matrix which is rich in clastic quartz and also contains chlorite, epidote and sericite of metamorphic origin. The pebbles are most strongly deformed in the south-west of the outcrop, close to the faulted western margin of the conglomerate, where they appear as somewhat elongated discs (Plate IIIB). Farther east the pebbles are almost cigar-shaped and strongly elongated along the regional lineation.

Structural Synopsis

Flinn (1958) has put forward an interpretation of the complex structure of Unst and Fetlar and the following summary is based on his work. The '*Basement*' is formed of the metamorphic rocks of the Valla Field and Lamb Hoga blocks, together with the migmatitic gneisses which underlie the whole of Yell and part of Shetland Mainland. The Valla Field and Lamb Hoga rocks have been tentatively correlated with the Scatsta Division of Mainland (Flinn and others 1972, fig. 1). The Saxa Vord Block of Unst, which is separated from the Valla Field Block by a major shear zone, could be either an early nappe of the Nappe Pile, or could have been emplaced separately at a much earlier date. The schists forming the greater part of this block bear a strong mineralogical resemblance to the Dunrossness Phyllites of Mainland. The *Nappe Pile* consists of two major nappes separated from each other and from the basement by schuppen or imbricate zones. The *Lower Schuppen Zone* extends along the central depression of Unst from Queyhouse southward to Belmont and includes Read's narrow southward extension of the Saxa Vord Block (Fig. 7a). This belt contains tectonic slices of phyllite, limestone and hornblende-schist which, as far as can be observed, dip at 45° to 90° to the east. The zone is not exposed in Fetlar, where it is apparently cut out by the Lamb Hoga Fault. The *Lower Nappe* is represented by the Main and Mu Ness serpentine blocks of Unst and by the Hamars Ness and Hesta Ness serpentine blocks in Fetlar. In the greater part of Unst the lower surface of the nappe dips steeply eastwards. In south-east Unst and in Fetlar, however, parts of the nappe reappear to the east of tectonically higher rocks, suggesting that here the nappe is folded into a south-south-west-plunging synform, (Fig. 7b). The *Middle Schuppen Zone* overlies the lower nappe. In Unst it is represented in the north by the small tectonic blocks of phyllitic, graphitic and hornblendic rocks around Norwick, and in the south by the Muness Phyllite. In Fetlar it forms all but one of the small blocks of phyllite, graphite-schist, metagabbro, migmatite, etc. exposed in the central section of the island. On the west and east sides of Vord Hill these are conspicuously overthrust by the Vord Hill Serpentine. Similarly, the schuppen-zone of Norwick in Unst is overthrust by the Clibberswick Serpentine. The *Upper Nappe* is best seen in Fetlar, where it is represented by the extensive outcrop of the Vord Hill Serpentine. This appears to occupy the core of a synform and is separated from the underlying rocks of the Middle Schuppen Zone by an inward-dipping thrust plane, exposed both in the east and west. Though the evidence is not conclusive, it is assumed that this thrust has a synclinal shape and is continuous below the entire serpentinite outcrop. On Unst the upper nappe is represented by the Clibberswick Block, which is lithologically identical with the Vord Hill Block. On Fetlar the remains of an *Upper Schuppen Zone* may be preserved. They form a small outcrop of hornblende-schist, serpentinite and phyllite which is exposed at Aith Ness on the south side of the island and which overlies the Vord Hill Serpentine. The Funzie Conglomerate appears to be an integral part of the Nappe Pile but its relationship to the latter's structural units is not clear.

The nappes appear to have been cut from an ultrabasic and basic layered intrusion which had suffered low-grade metamorphism prior to the formation of the nappes. The lower nappe is in places several kilometres thick and the upper nappe has a thickness of over $1\frac{1}{2}$ km. As the Middle Schuppen Zone contains low-grade metasediments and hornblende-schists which do not occur in either the basement or the nappes, it is likely that the upper nappe, at least, has moved a considerable distance to its present position. The phyllites of the Middle Schuppen Zone also contain much material which appears to be derived by subaerial erosion and re-deposition from the rocks which now form the nappes. This suggests that the lower nappe was exposed at the surface and being actively eroded before the upper nappe was emplaced, and that both nappes were formed at a very high tectonic level. After the emplacement of the nappes the whole nappe-pile was folded into a great synform plunging to the south-south-west with limbs dipping at 45° or more.

Garson and Plant (1973) have suggested that the ultrabasic rocks of the Nappe Pile are the remnants of an ophiolite complex and that the thrusts bounding the individual nappes have acted as subduction shear planes. They have correlated the Unst–Fetlar subduction zone with that extending along the Highland Boundary Fault of the Scottish Mainland. Plant and others (in press) have carried out a detailed investigation into the distribution of trace quantities of metal ores within the complex.

REFERENCES

Amin 1952, 1954; Curtis and Brown 1969; Fernando 1941; Flinn 1952, 1956, 1958, 1959, 1970b; Garson and Plant 1973; Heddle 1878; Miller and Flinn 1966; Phemister 1964; Phillips 1927; Plant and others (in press); Read 1933, 1934a, b, c; 1936a, b; Snelling 1958; Summ Prog. 1930, 31, 32, 58.

D

5. BASEMENT COMPLEX OF ORKNEY

The crystalline basement rocks of Orkney form a number of small inliers near Stromness, all of which lie along a north-north-west trending belt extending from Yesnaby to Graemsay (Fig. 16). The largest outcrop forms the hilly ground just north of Stromness, which reaches a height of 96 m O D.

A high proportion of the basement is made up of coarse pink or greyish poorly foliated granite which locally grades into granite-gneiss. The granite contains enclaves of biotite-gneiss and smaller masses of siliceous, micaceous and hornblendic schist. Both the granite and included country rock are veined by fine-grained pink granite and by pegmatite. On Graemsay the included bodies of schist, which near the northern end of the island include hornblende- and pyroxene-schists, reach a size of several hundreds of metres. Within the Yesnaby inlier there are large masses of siliceous schist, and, on the shore just north of Harra Ebb [217 150], hornblende- and biotite-schist.

No detailed structural or petrographic work has been done on the basement complex. Wilson (Wilson and others 1935, p. 15) suggested that the granite forms part of a deep-seated mass of which only the top is seen. He compared the metamorphic rocks with the Lewisian inliers found in the Moine of the Altnaharra district. In the southern inliers the coarse granite is generally slightly foliated and appears to pass in places into granite-gneiss, suggesting it forms part of the suite of older Caledonian granites.

A small outcrop of pale grey flow-banded porphyritic felsite can be seen on the promontory at the west side of the Bay of Navershaw (Fig. 16) 1·5 km E of Stromness. This is unconformably overlain by a thin breccia which contains clasts of the felsite and passes laterally into Stromness Flags. It is not known if the felsite is itself underlain by Stromness Flags or if it forms part of the basement on which the flagstones rest. Wilson (in Wilson and others 1935, p. 49) regarded it as a pre-Middle Old Red Sandstone rock, but Michie (*in* Gallagher and others 1971, p. B 166) has suggested that it may be an ignimbrite of Lower Old Red Sandstone age.

The pre-Old Red Sandstone Land Surface. Prior to the deposition of the Old Red Sandstone the basement inliers appear to have formed a range of hills which was elongated in a north-north-westerly direction. The present topography of the inliers appears to be much the same as it was at the time of their burial, and small patches of sediment can still be found on the hill tops of Yesnaby and Graemsay. The original hills may not have been more than 100 m high and the original average gradients of their slopes did not normally exceed 13°, but there is ample evidence for the presence of small cliffs, knolls and gullies. The granite and metamorphic rock underlying the exhumed surface is everywhere fresh and there is no sign of any deeply weathered profile.

REFERENCES
Fannin 1969, 1970; Gallagher and others 1971; Steavenson 1928a; Wilson and others 1935.

6. PLUTONIC COMPLEXES OF SHETLAND

Introduction

The Shetland Islands contain two groups of Caledonian plutonic complexes (Fig. 8). The earlier of these is intruded into the metamorphic rocks east of the Walls Boundary Fault and includes the Graven and Brae complexes of Delting, the Spiggie Complex of central and south Mainland and the very small areas of granite at Channerwick and Cunningsburgh in south-eastern Mainland. Some of these complexes are in places weakly foliated, indicating that they were emplaced before Caledonian tectonic activity had ceased. The second group crops out just west of the Walls Boundary Fault and forms a north–south trending series of possibly interconnected complexes which extend from North Roe to the southern side of the Walls Peninsula. A possible representative of this group on the east side of the Walls Boundary Fault is the granite that may lie just off the south-west shore of Fair Isle (p. 96). These complexes are not foliated and are thus essentially post-tectonic intrusions. The Sandsting Plutonic Complex, however, cuts Middle Old Red Sandstone sediments, which were intensely folded during or *after* the intrusion. This suggests that the emplacement of at least some of the complexes may have been connected with the final localised movements of the Caledonian Orogeny. Swarms of acid and basic dykes (p. 94) are centred on some of the western complexes.

Eastern Late-Orogenic Complexes

Graven Complex

This intrusive complex occupies an area which extends from the Walls Boundary Fault in Northmaven eastward *via* Sullom Voe, northern Delting and the adjoining islands in Yell Sound to the Nesting Fault. It reappears further south in North Nesting where it is exposed along the south shore of Dury Voe, having apparently been displaced dextrally for 14 to 16 km by the Nesting Fault (Fig. 8).

The Graven Complex is not a single massive intrusion. It could be more accurately described as two superimposed vein complexes which cut the metamorphic country rocks and now occupy about half of the affected area. This means that within the complex there are now many large and small enclaves of country rock, most of which maintain their original position and orientation so that the original structure of the metamorphic rocks can still be determined. The first major phase of intrusion was the emplacement of a plexus of coarse granite-pegmatite veins, pods, dykes and irregular larger bodies which are now separated by screens of variably permeated country rock. This was followed by the intrusion of lamprophyre and porphyrite dykes (p. 95), which extend beyond the limits of the complex. The second major phase was the intrusion of a network of medium- to fine-grained hybrid plutonic rocks which Flinn has collectively termed the *Inclusion Granite*. These consist of a continuously variable mixture of diorite, monzonite, granodiorite and granite and are characterised by the abundance of small rounded cognate xenoliths of almost pure hornblende rock. The most granitic portion of the complex forms the *Stava Ness Granite*. The Inclusion Granite is also full of orientated angular inclusions and larger enclaves of country rock. The effects of thermal metamorphism by the intrusive rocks of the complex are most marked in the xenoliths. These contain abundant sillimanite in the pelitic rocks and diopside in the calcareous ones.

Brae Complex

The Brae Complex is the most northerly and largest of a series of ultrabasic to acid plutonic complexes which extend as far south as Whitelaw Hill [356 540], 2 km SE of Aith. These complexes have a family similarity and all have amphibole as their most characteristic mineral. Their emplacement may well represent an early phase in the igneous activity which culminated in the formation of the Graven Complex.

The Brae Complex itself (Fig. 8) occupies the south-western peninsula of Delting which is bounded in the west by Busta Voe and in the south by Olna Firth. Unlike the Graven Complex it is a compact, composite stock-like intrusion. It is largely composed of a two-pyroxene-diorite with andesine, potash-feldspar and biotite, which has been extensively altered to a dioritic rock composed of the original andesine and new green secondary amphibole, biotite and quartz. Scattered throughout the complex are masses of ultra-basic rock, the largest of which lies in the south-west corner. These appear originally to have varied from peridotite to pyroxenite (with both ortho-pyroxene and clinopyroxene) and dunite, although most of the olivine is now serpentinised and the pyroxenes are commonly altered to amphibole. In good coastal exposures it is possible to see that these rocks are composed of variably sized and irregularly shaped fragments of pyroxenite and pyroxene-rich peridotite set in a sparse matrix of serpentinised dunite and olivine-rich peridotite.

In two places along the south coast there are small xenolithic areas. The xenoliths include altered metasedimentary and ultrabasic rocks and they are associated with two-pyroxene gabbro. These rocks are characterised by the presence of reddish-brown hornblende; they are considerably richer in iron than similar rocks in the rest of the complex, and Gill (1965) considers them to be an early phase of emplacement. However, all the pyroxenites and pyroxene-rich peridotites almost invariably contain traces of the reddish-brown hornblende, suggesting that *all* the ultrabasic rocks probably belong to the early phase of emplacement.

The original pyroxene-rich rocks (cumulates?) appear to have been mobilised by brecciation and injection with olivine-rich peridotite and trans-ported to a higher level in the crust. There, together with the iron-rich xenoliths and the gabbro, they developed reddish-brown hornblende. Large masses of these rocks were then transported as xenoliths in the two-pyroxene diorite to their present positions in the Brae Complex. In some of the complexes to the south of Brae the ultrabasic rocks appear to have remained in

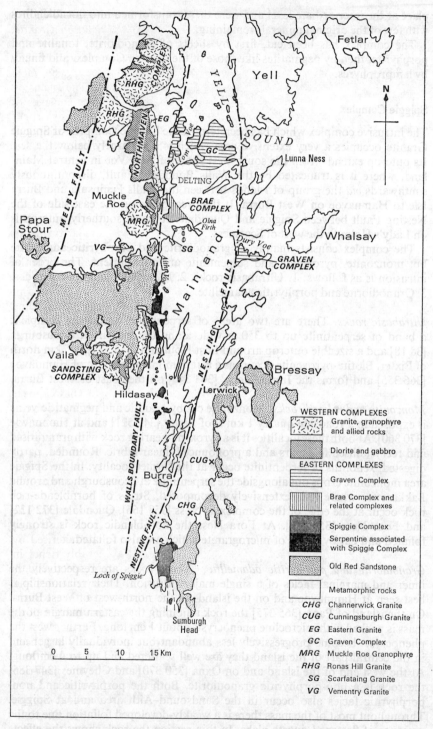

Granite, granophyre
and allied rocks

Diorite and gabbro

EASTERN COMPLEXES

Graven Complex

Brae Complex and
related complexes

Spiggie Complex

Serpentine associated
with Spiggie Complex

Old Red Sandstone

Metamorphic rocks

CHG   Channerwick Granite
CUG   Cunningsburgh Granite
EG   Eastern Granite
GC   Graven Complex
MRG   Muckle Roe Granophyre
RHG   Ronas Hill Granite
SG   Scarfataing Granite
VG   Vementry Granite

FIG. 8. *The plutonic complexes of Shetland Mainland*

place at the intermediate level and were there transformed into hornblendites with few of the original minerals remaining.

The complex has been cut, first by sheets of granodiorite, tonalite and porphyrite, then by pegmatites like those of the Graven Complex, and finally by lamprophyres.

Spiggie Complex

The intrusive complex which has been termed the Spiggie Complex or Spiggie Granite occupies a very extensive area which is now largely below the sea. Its outcrop extends from the south-west shore of Aith Voe in central Mainland, where it is truncated by the Walls Boundary Fault, discontinuously southwards *via* the group of islands between the Walls Peninsula and Burra Isle to Hamnavoe on West Burra. It then reappears on the east side of the Nesting Fault between Spiggie and Quendale. Its most southerly outcrop is on Lady's Holm to the west of Scatsness.

The complex consists mainly of granodiorite and porphyritic adamellite but monzonite, pyroxenite and serpentinite are also present. The order of intrusion is as follows: 1. Ultrabasic rocks, 2. Monzonite and related rocks, 3. Granodiorite and porphyritic adamellite.

Ultrabasic rocks. There are two areas of serpentinite within the complex: a band of serpentinite up to 350 m wide along the margin at Scousburgh [38 18] and a sizeable outcrop around East Houlland [345 535], 1·5 km north of Bixter. Biotite-pyroxenite occurs as a raft in monzonite south of Hamnavoe [366 355] and forms the Inner Skerry [363 340] off the coast of West Burra.

Monzonite and related rocks. Monzonite cut by granitic and pegmatitic veins is exposed in a roadside quarry 1 km E of Bixter [342 521] and at Hamnavoe [370 360]. At both these localities it is a pyroxene-bearing rock with granulitised and recrystallised feldspars and a pronounced linear fabric. Rounded, partly digested, xenoliths of serpentinite occur at the former locality. In the Spiggie area monzonite crops out alongside the serpentinite at Scousburgh and around Bakkasetter where it is extensively decomposed. Strips of hornblende-rich rock occur at the edge of the complex at Noss [358 166], Quendale [372 132] and Fora Ness [350 450]. At Fora Ness the hornblendic rock is strongly foliated and cut by dykes of microgranite which are also foliated.

Granodiorite and porphyritic adamellite. These rocks are respectively the inner and marginal facies of a single major intrusion. Their relationship is best seen at Hamnavoe and on the islands to the north-west of West Burra. On the island of Papa [365 375] the rock adjoining the eastern margin of the mass is packed with microcline phenocrysts about 1 cm long. Farther west the phenocrysts become progressively less abundant but individually larger and in the central part of the island they are well scattered and up to 4 cm long. At the west end of the island and on Oxna [350 370] and Cheynies [347 386] the rock is a non-porphyritic granodiorite. Both the porphyritic and non-porphyritic facies also occur in the Sandsound–Aith area and at Spiggie. Throughout most of the mass there is a weakly-developed foliation due to the presence of flattened quartz blebs. In thin section the rock shows the effects of incipient granulitisation and recrystallisation and contains abundant secondary epidote. Parts of the mass are cut by veins of pegmatite up to 1 m thick.

In south Mainland the Spiggie Granite is unconformably overlain by Old Red Sandstone sediments. Where it is in contact with the metamorphic rocks it has generally produced only a narrow thermal aureole, but at its contact with the Dunrossness Phyllitic Group the thermal effects are extensive and spectacular. At Scousburgh, for instance, the newly formed minerals include: garnet, andalusite, sillimanite, kyanite, chloritoid and staurolite. The distribution and textural relations of these minerals suggest that the thermal metamorphism took place in at least two stages separated by a period of deformation during which a strain-slip cleavage developed (see Flinn 1967a, pp. 270–73).

Channerwick and Cunningsburgh Granites

The Channerwick Granite consists of a number of interconnected dyke-like bodies up to 25 m thick with sill-like offshoots, which are exposed close to the main road at Channerwick, 18 km SSW of Lerwick. It consists of albite, muscovite and quartz and is associated with sheets of quartz-porphyry. This small outcrop is remarkable for the extent of its thermal aureole which has a diameter of about 900 m. The aureole is conspicuous because of the presence of white and dark spots, consisting respectively of shimmer aggregate and chlorite, in the surrounding Clift Hills Phyllitic Group. Of even greater interest is the presence of newly formed chloritoid in Dunrossness phyllite within the outer zone of the aureole, more than 500 m from the granite outcrop.

The outcrop of the Cunningsburgh Granite is even smaller, but as it is intruded into Dunrossness phyllites which are highly sensitive to thermal metamorphism, its aureole is extensive and consists of an inner zone containing porphyroblasts of staurolite, a middle zone in which chloritoid has partially replaced staurolite and an outer zone of completely recrystallised chloritoid.

Western Post-Orogenic Complexes

Northmaven Complex

The outcrop of the Northmaven Complex (Fig. 4) extends from the Beorgs of Uyea, close to the northern coast of Shetland Mainland, southwards *via* the great mass of Ronas Hill and the rugged terrain which forms most of Northmaven, to the island of Muckle Roe and beyond into Vementry island. The complex includes various types of granite, granophyre, diorite and gabbro, and a few small outcrops of ultrabasic rock as well as some altered basaltic rocks.

The emplacement of the plutonic rocks may have taken place in the following three stages: (1) Intrusion of basic magma: (2) Intrusion of acid magma and hybridisation: (3) Late intrusion of granophyre. The plutonic activity was preceded by the formation of early minor hypabyssal intrusions and perhaps also lavas. It was followed by the intrusion of great swarms of acid, intermediate and basic dykes (Chapter 9). At a late stage the rocks were affected by hydrothermal alteration which uralitised the basic plutonic and hypabyssal rocks. Even later there was a period of localised mineralisation leading to the formation of scapolite and zeolite. Scapolite is seen in veins along the eastern margin of the complex at Mavis Grind and further north.

Early hypabyssal intrusions and lavas. Intrusive rocks which are older than the plutonic rocks have been identified within the Ronas Hill Granite and on the Isle of Egilsay at the mouth of Mangaster Voe, where metamorphic rock cut by a dyke of porphyrite is enclosed in and thermally altered by diorite. Small masses of dolerite and basalt altered by the surrounding diorite or gabbro are also found in the Busta Peninsula and in the eastern coastal area of Muckle Roe. The coastal stretch between Gunnister Voe and Mangaster Voe, particularly around Wilson's Noup, contains many irregular xenoliths of altered fine-grained basic igneous rock enclosed and partly assimilated by a network of veins of diorite and granodiorite (Plate VB). These xenoliths may be the remnants of a series of early lava flows whose composition may perhaps be representative of the undifferentiated parent magma of the complex.

Ultrabasic Rocks. Blocks of ultrabasic rock ranging from harrisite to lherzolite have been recorded at two localities just west of Clothister Hill within the dioritic portion of the complex (Fig. 4). These rocks are composed mainly of orthopyroxene, clinopyroxene, calcic plagioclase and olivine. It is not known if they are derived from small discrete bosses or from detached enclaves within the diorite. A different type of ultrabasic rock, hornblende-hypersthenite, crops out near the head of Ronas Voe, where it forms a small body entirely enclosed in metamorphic rock. The nature of its contact with the surrounding rock is, however, not seen.

Diorite and Gabbro. Plutonic rocks of intermediate and basic composition occupy three main areas within the complex (Fig. 4). The two smaller outcrops are in North Roe, where they are largely enclosed in the Ronas Hill Granite; the largest extends from the shores of Ronas Voe southwards to the east coast of Muckle Roe. Diorites are by far the most abundant rock types in these outcrops and in many areas they are riddled by granite and granophyre veins, giving rise to a very large net-vein complex. Gabbro occurs in small irregular areas throughout the diorite, and in these the rocks of different composition are closely intermingled, with gabbros passing laterally within a short distance into dioritic or more acid types. The greatest concentrations of gabbro occur in the southern part of the area, around Mangaster Voe and Mavis Grind and in the Busta Peninsula, but there is also one large outcrop along the north shore of Clubbi Shuns in North Roe.

The *gabbros*, though variable in grain-size and composition, are generally medium- to coarse-grained. They grade from true gabbro into hornblende-biotite-diorite with minor quartz and potash-feldspar. Both the augite and hornblende of these rocks are completely or peripherally altered to uralitic amphibole, which also occupies larger pockets resembling vesicles. Some of the fine-grained basic rocks are uralitised dolerites or basalts. These are most abundant near Mavis Grind and on the east shore of Muckle Roe.

The *diorites* are medium-grained black and white speckled rocks with a

wide range in texture, grain-size and composition. They range from granodiorite to fine-grained hornblendic meladiorite. These variations do not have
any relation to their position in the complex. The diorite outcrops also contain
many small acid bodies with irregular outcrops and in many areas the latter
form a network of veins. The junctions between acid and more basic material
may either be sharp and angular, suggesting that the latter had consolidated
before veining took place, or highly undulating, which might point to the
existence of adjoining fluid magmas. Other contacts again are diffuse due to
metasomatic feldspathisation and coarsening within the basic xenolith.

Granite and Granophyre. The granitic components of the Northmaven
Plutonic Complex belong to a number of separate bodies which may have
been emplaced during distinct phases in the development of the complex.

In the north the earliest granitic intrusion is a near-vertical dyke-like body
(? ring-dyke) of xenolithic granite which can be traced for a distance of
5·5 km along the south-eastern border of the Ronas Hill Granite (Fig. 4).
It is 60 m wide north of the Brig of Colla Firth, but widens south-west of the
Brig to form a zone of xenolithic granite and schist about 250 m wide. The
xenoliths are angular and consist of siliceous and micaceous schist with
occasional fragments of hornblende-schist. Along a considerable part of its
outcrop a screen of metamorphic rock intervenes between the dyke and the
granite to the west.

The main granitic body in North Roe is the *Ronas Hill Granite* (Plate V A).
In the west it is separated from the volcanic rocks of Esha Ness by the
probable northward continuation of the Melby Fault (p. 52); in the north
and east it has a near-vertical contact with the metamorphic country rock
and the diorite. The 'granite' is in fact a deep red leucocratic granophyre.
It contains cavities (druses) which are generally about 1 mm in diameter, and
which often contain stilpnomelane and are in places lined with crystals of
milky quartz.

A long narrow dyke-like outcrop of red leucocratic granite 150 m to
1·6 km wide, called the *Eastern Granite*, extends along the eastern margin of
the complex from the head of Ronas Voe southwards for 14·5 km past Mavis
Grind to Busta Voe. The eastern margin of this mass is near-vertical and
appears to be controlled by the existence of an early dislocation along the line
of the Walls Boundary Fault zone (p. 20). On its western margin, however,
the contact with the gabbro-diorite takes the form of a plexus of large and
small sheets, dykes and veins of granite in the diorite. These relationships
indicate that the Eastern Granite was intruded under conditions which
permitted easy penetration of the diorite.

The most southerly outcrops of granite within the Northmaven Plutonic
Complex occur near Scarfataing in the south-east of Muckle Roe and in the
north-eastern part of the island of Vementry (Fig. 8). The *Scarfataing
Granite* contains large enclaves of gneiss, which are cut by dykes of ophitic
dolerite but not by diorite, suggesting that it was an early intrusion which
may even have preceded the emplacement of the diorite. It appears to have
been emplaced as a series of sheets which are now vertical. In many exposures
both the granite and the included gneiss are sheared. The *Vementry Granite*
consists of two lithological types, an outer coarse-grained pink leucocratic
quartz-rich granite and an inner granite which consists of phenocrysts of

feldspar and quartz set in a darker, relatively fine-grained matrix. The junction with the adjoining metamorphic rock is straight and sharp and where seen it is inclined at 40° to 50° away from the granite. This granite is of particular interest in that it forms the focus of a large number of acid dykes which appear to radiate from its centre (Fig. 24).

The youngest rock in this southern part of the Northmaven Plutonic Complex is the *Muckle Roe Granophyre* which forms most of Muckle Roe and appears to have originated as a roughly circular stock. The granophyre is closely jointed and gives rise to rugged topography with extensive inland scree slopes and impressive coastal cliffs. It is pink in colour, locally tinged with yellow, and consists of blebs of quartz set in a stony base in which feldspar cleavage faces are sometimes seen. Drusy cavities are common and these may be lined with well-shaped quartz crystals.

Intrusive history and shape of complex. The Northmaven Plutonic Complex appears to have been emplaced during a fairly prolonged period of magmatic activity in the Devonian Period. The complex may be younger than the volcanic rocks of Esha Ness (p. 57), and as is suggested by the radiometric (Rb:Sr) date of 358 ±8 m.y. for the Ronas Hill Granite (Miller and Flinn 1966) its period of emplacement may span the Middle–Upper Devonian boundary.

The main stages in the evolution of the complex are as follows:

1. Emplacement of early minor intrusions and ? lavas.
2. Emplacement of members of the plutonic complex in the following order:
   (a) Early granite ring-dykes (i.e. Colla Firth, Scarfataing and ? Vementry granites)
   (b) Diorites and gabbros
   (c) Eastern Granite (in part contemporaneous with (b))
   (d) Circular or oval granophyric stocks (i.e. Ronas Hill Granite and Muckle Roe Granophyre).
3. Emplacement of swarms of late minor intrusions.

Phemister (1950, p. 360) has shown that the junction of the Ronas Hill Granite with the adjoining gneiss is everywhere sharp and steep, and he believes that the intrusion of the granite was preceded by the formation of an arcuate zone of fracture and brecciation, along part of which an early ring-dyke was intruded. The main body of magma then punched out a clean plug from the loosened country rock within this zone.

The diorite-gabbro component of the complex is believed to have been derived from a basic magma which was rich in water and whose emplacement was slow and tranquil, taking place virtually under hydrostatic conditions. During its emplacement there was a continuous acidification of the gabbroic rock by reaction, in part with the hydrous and quartzo-feldspathic fluids derived from the differentiation of the basic magma and in part with granitic magma derived from the limited accession of material from adjacent granitic magma reservoirs. Granitic magma was subsequently intruded, first quiescently as dykes and sheets within the diorite and as the irregular dyke-like Eastern Granite, and later forcefully by pushing its way upward to produce the stocks which now form the Ronas Hill Granite and Muckle Roe Granophyre.

A. Cliffs of Ronas Hill Granite, Heads of Grocken, Shetland north Mainland, looking east towards Hillswick peninsula (D1653)

Plate V

B. Net vein complex of basic and granitic rock, Northmaven Complex. Wilson's Noup, Northmaven, Shetland Mainland (D1345)

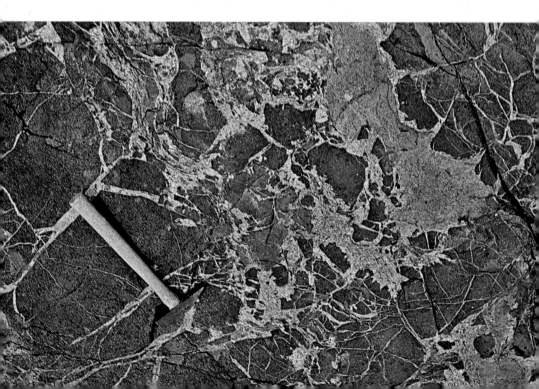

Sandsting Complex

The Sandsting Complex forms the south-eastern part of the Walls Peninsula and is intruded into the sedimentary rocks of the Middle Old Red Sandstone Walls Formation (Fig. 9). As in the Northmaven Plutonic Complex the rocks range from granitic to ultrabasic, but here gabbro is a very minor constituent and dioritic rocks form a smaller proportion of the total outcrop. As some of the outcrop of this complex may now be below sea level and only a small part of its margin is exposed, it is difficult to ascertain the shape and structure of the complex. In the western part of the area the junction with the sediments is inclined at 40° to 70° to the north, but in the east granite forms a series of near-vertical north-north-west trending sills which thin out northwards. The contact is in all areas roughly sub-parallel to the bedding of the sediments.

As in Northmaven a number of phases in the evolution of the complex can be recognised. These are as follows:

1. Intrusion of early basic and sub-basic dykes.
2. Emplacement of plutonic components in the following sequence:
   (a) Diorite and gabbro (i.e. basic magma)
   (b) Several pulses of acid magma producing in turn granodiorite, biotite-granite, granophyre and porphyritic microadamellite.
3. Formation of acid and sub-basic minor intrusions.
4. Scapolitisation along active shear belts.

Diorite, gabbro and ultrabasic rock. Dioritic rocks form a number of irregular outcrops between the mouth of Gruting Voe and Garderhouse. These are almost completely surrounded by granitic rocks. Two of the outcrops, at Culswick and at Wester Skeld, contain large and small enclaves of horn-felsed sediment, the largest of which is over 800 m long. The dioritic rocks range in composition from melamicrodiorite, through hornblende-diorite to quartz-biotite-diorite and granodiorite. On Hestinsetter Hill there is an outcrop of vertically banded diorite full of minute ovoids which consist of pyroxene-monzonite with conspicuous euhedral crystals of sphene.

Gabbro forms only two small outcrops on the east shore of Skelda Voe and near Garderhouse. As in Northmaven it has primary hornblende as well as augite and is strongly uralitised and saussuritised. The presence of a small outcrop of ultrabasic rock is inferred from a concentration of boulders near Stump Farm, just east of Gossa Water.

The diorite is in places veined by leucocratic quartz-diorite and pegmatite with abundant epidote. All these veins are considered to be the products of differentiation of the dioritic magma. In addition there are areas where the vein material is derived from the invading granitic magma. In these areas the margins of the xenoliths can be either angular or highly undulating, indicating that the diorite had reached various stages of consolidation when the granitic magma was intruded.

Granitic rocks. In the Culswick area lenticular east–west trending belts of granodiorite, porphyritic microgranite and porphyritic microadamellite adjoin the diorite (Fig. 9). Farther east the granitic outcrop is formed by leucocratic biotite-granite, which completely encloses the central and eastern

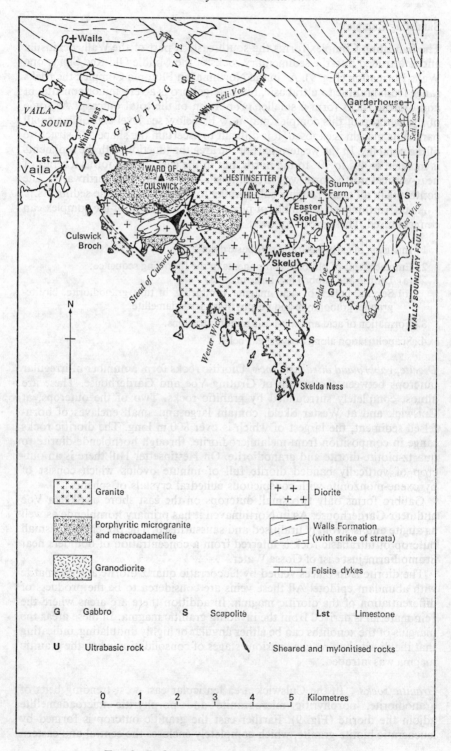

FIG. 9. *Geological map of the Sandsting Complex*

diorite outcrops and also forms the granite sills in the north-east of the complex. It passes into a coarse graphic granite in the eastern sills and this becomes progressively more fine-grained and granophyric in texture towards Bixter Voe.

Scapolitised shear belts. In the area around Wester Wick and in the Skelda Ness Peninsula the granite contains a number of lenticular, near-vertical north-north-west trending belts of intensely sheared and locally mylonitised rock. Sodic scapolite occurs both as a replacement product and as a vein-filling mineral in these belts; it also forms narrow veins in basic minor intrusions within the complex and in the sediments adjoining the granite. The scapolite was probably introduced by late hydrothermal solutions along active shear belts, joints, and other lines of weakness at some time after the emplacement of the granite. These scapolitised shear belts do not appear to be connected with the Walls Boundary Fault.

Intrusive history and shape of complex. No part of the Sandsting Complex has either the shape or the marginal relationships which could suggest that it forms part of a vertical or steep-sided stock at the present level of erosion. It is more likely that the complex originated as a sheet which was emplaced in several pulses of magmatic activity. After an early hypabyssal phase, when a few basic dykes were formed, the first major intrusion appears to have been by basic or sub-basic magma. This gave rise to a sill-like sheet of diorite which enclosed several masses of sediment. Before its consolidation was complete granitic magma was intruded in several pulses along or close to both the upper and lower margins of the diorite and also beyond its limits. Locally the diorite sheet was cut and transgressed by the granite. During this phase a number of lenticular granodioritic and granitic sheets were formed, the latter being the most extensive.

Though in the eastern part of the complex both the top and base of the granite sheet are seen, west of Skelda Voe there is no exposure of the base. The possibility that the western part of the outcrop is part of the upper portion of a stock cannot therefore be ruled out.

REFERENCES

Finlay 1930; Flinn 1954, 1967a; Gill 1965; Miller and Flinn 1966; Mykura and Phemister 1976; Mykura and Young 1969; Phemister and others 1950; Pringle 1970; Summ. Prog. 1933.

7. OLD RED SANDSTONE OF SHETLAND

The outcrops of Old Red Sandstone in the Shetland Islands fall into three groups separated from each other by two large, probably transcurrent, faults: the Melby Fault in the west and the Walls Boundary Fault in the centre of Shetland Mainland (Fig. 10).

Western Outcrops

West of the Melby Fault Middle Old Red Sandstone sediments and volcanic rocks crop out in the north-west corner of the Walls Peninsula where they are known as the Melby Formation, on the island of Papa Stour, and in Esha Ness on the west side of north Mainland. Old Red Sandstone sediments of uncertain age form the greater part of the island of Foula.

Walls Peninsula

Melby Formation

The rocks ascribed to the Melby Formation consist of buff and red sandstones, pebbly sandstones and sandy siltstones intercalated at the top with two thick flows of silicified rhyolite or ignimbrite. Near the base of the exposed sequence there are two beds of pale grey siltstone and shale with bands and nodules rich in carbonate. These are known as the Melby Fish Beds and they contain both fish and plant remains. The fish fauna includes *Cheiracanthus sp.*, *Coccosteus cuspidatus* Miller *ex* Agassiz, *Dipterus valenciennesi* Sedgwick and Murchison, *Glyptolepis* cf. *leptopterus* Agassiz, *Gyroptychius agassizi* (Traill), *Homostius milleri* Traquair, *Mesacanthus sp.* and *Pterichthyodes sp.* This assemblage is very similar to that of the Sandwick Fish Bed fauna of Orkney (p. 73), and the fish beds of the two areas have been confidently correlated. The lower part of the Melby Formation may thus be the stratigraphical equivalent of the Stromness Flags and may therefore range from topmost Eifelian to lower Givetian (Table II).

The sediments below the rhyolites can be divided into two major groups:

1. The lower group contains the two fish beds set in a sequence of red and buff cross-bedded sandstones. The sandstones are of fluvial origin and were deposited by currents from the west or west-north-west. It is possible that this part of the Melby Formation was laid down close to the north-western margin of the shallow but extensive Orcadian Basin. For most of the time the area was part of an alluvial plain or fan but on two occasions it became submerged under the waters of a temporarily transgressing lake. The periods of submergence are now represented by the fish beds. The lake in which one or both of these beds was laid down may have extended to Orkney and Caithness, where it gave rise to the Sandwick Fish Bed and Achanarras Limestone respectively.

2. The upper group contains thick beds of pink feldspathic sandstone with clasts of rhyolite and basalt and many plant fragments together with thick beds of purplish bioturbated sandy siltstone. These strata appear to have been deposited by currents from the east-north-east. Prior to and during the deposition of these beds the earlier topography and drainage pattern of the area were probably drastically altered by outpourings of volcanic rocks, which led to the formation of volcanic hills to the north-east of the present outcrop. The thick beds of purple sandy siltstone are of a type normally laid down in relatively quiescent waters, suggesting that some of the drainage may have been ponded to form local lakes.

Papa Stour

The island of Papa Stour (Fig. 2) consists of two thick flows of rhyolite with intercalated tuffs, underlain by basalts and sandstones. It is possible that the rhyolites are the equivalents of the rhyolitic flows in the Melby Formation and that the basalts are represented farther south by thin basalt flows exposed on the Holm of Melby [193 585]. The rocks of Papa Stour also resemble the lower part of the Esha Ness volcanic sequence (p. 57), with which they have been equated.

The sequence and approximate thickness of the volcanic and sedimentary rocks is as follows:

	Thickness (m)
Upper rhyolite	85+
Rhyolitic tuff and agglomerate . . .	2·5–24
Lower rhyolite	0–40
Rhyolitic tuff and tuffaceous sandstone in west, passing into sandstone with tuffaceous bands in east . .	0–30+
Basalt (up to 4 flows exposed)	24+

The basalts generally have thick scoriaceous upper zones, with vesicles filled with chalcedony, calcite, barytes and zeolite. Agates with barytes cores are present in some exposures and along the south coast fluorspar, heulandite and stilbite have been recorded. In the east of the island the highest basalt is truncated by an uneven erosion surface which contains small channels filled with the overlying sandstone, but in the south-west it is overlain by basaltic rubble. The rhyolitic tuff above this rubble not only rests on an undulating floor but is itself truncated by an irregular erosion surface. The tuff passes eastward into a much thicker sequence of sandstone with tuffaceous bands and with some beds of red bioturbated sandy siltstone. The two flows of rhyolite form the impressive orange-red sea cliffs of Papa Stour, which on the west coast are cut by long interconnected sea caves eroded along joint planes and faults. The rhyolite is strongly banded, generally sparsely porphyritic and, in many areas, full of minute spherulites. In places along the north-west and south coasts it contains larger closely packed near-spherical bodies which are up to 4 cm in diameter. These are known as lithophysae. The tuffaceous deposit between the two rhyolite flows is well exposed on the north-west coast of the island, where it rests on an uneven strongly eroded floor of rhyolite (Plate VIII B).

The Papa Stour rhyolites are completely devitrified and recrystallised and little of their original texture remains. All remaining textural evidence, such as the steeply inclined flow banding with aligned spherulites and lithophysae,

TABLE II   POSSIBLE   CORRELATION   OF   THE   OLD   RED

STAGE	FISH FAUNA (Caithness, Orkney & Melby)	CAITHNESS			ORKNEY		
? UPPER O.R.S.		Top not exposed **DUNNET HEAD SANDSTONE**			Top not exposed **HOY SANDSTONE** **1000+**		
		(unconformity)			**HOY VOLCANICS**	0–90 BASALT 0–15 TUFF	
		Unconformity			Gentle folding and some major faults		
GIVETIAN	*Tristichopterus alatus* *Microbrachius dicki* *Pentlandia macroptera* *Watsonosteus fletti*	**JOHN O'GROATS SANDSTONE GROUP 627+**			EDAY BEDS	UPPER EDAY SST 320+	
						EDAY MARLS 100	
						MIDDLE EDAY SST 100–500	
						EDAY FLAGS ?10–150	LAVAS
						LOWER EDAY SST 220–250	
						PASSAGE BEDS	
GIVETIAN	*Thursius pholidotus* *Millerosteus minor* *'Estheria' membrancea*	UPPER CAITHNESS FLAGSTONE GROUP 1500+	MEY SUBGROUP 553		**ROUSAY FLAGS** **1500+**		
			HAM-SKARFSKERRY S/G 750				
	Dickosteus threiplandi		LATHERON S/G 175+		**UPPER STROMNESS FLAGS** **330+**		
		ACHANARRAS & NIANDT LST			*SANDWICK FISH BED*		
COUVINIAN (= EIFELIAN)	*Coccosteus cuspidatus*	LOWER CAITHNESS FLAGSTONE GROUP 2350+	ROBBERY HEAD S/G 155+		**LOWER STROMNESS FLAGS** **215+**		
			LYBSTER S/G 870+				
	Thursius macrolepidotus		HILLHEAD RED BED S/G 160		Gentle folding giving rise to angular unconformity of 10°		
			CLYTH S/G 1150	HELMAN HEAD BEDS =BERRIEDALE FLAGS & SANDSTONE BADBEA BRECCIA			
		Local minor warping					
? EMSIAN AND SIEGENIAN		**SARCLET or BASEMENT GROUP (434 + at Sarclet dome)**			WAREBETH RED BED FORM	? HARRA EBB FORM	YESNABY SST FORM

CO Conglomerate   FORM Formation   LST Limestone

Thicknesses are given in metres

SANDSTONE   OF   SHETLAND,   ORKNEY   AND   CAITHNESS

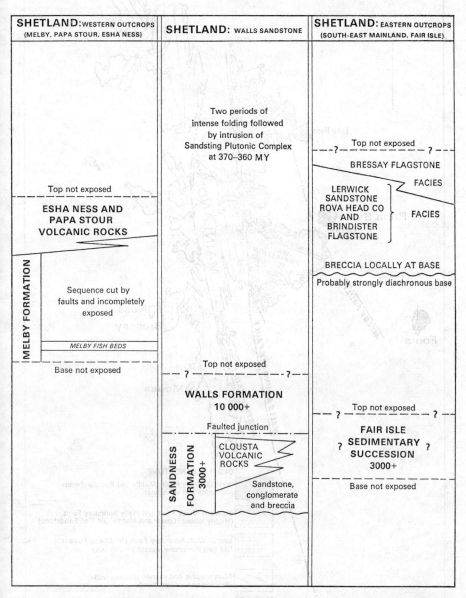

SHETLAND: WESTERN OUTCROPS (MELBY, PAPA STOUR, ESHA NESS)	SHETLAND: WALLS SANDSTONE	SHETLAND: EASTERN OUTCROPS (SOUTH-EAST MAINLAND, FAIR ISLE)

MELBY FORMATION

Top not exposed

ESHA NESS AND PAPA STOUR VOLCANIC ROCKS

Sequence cut by faults and incompletely exposed

MELBY FISH BEDS

Base not exposed

Two periods of intense folding followed by intrusion of Sandsting Plutonic Complex at 370–360 MY

Top not exposed
? — — — — — — — — — ?

WALLS FORMATION 10 000+

Faulted junction

SANDNESS FORMATION 3000+

CLOUSTA VOLCANIC ROCKS

Sandstone, conglomerate and breccia

? — Top not exposed — ?

BRESSAY FLAGSTONE
FACIES

LERWICK SANDSTONE ROVA HEAD CO AND BRINDISTER FLAGSTONE
} FACIES

BRECCIA LOCALLY AT BASE
Probably strongly diachronous base

Top not exposed
? — — — — — — — ?

? **FAIR ISLE SEDIMENTARY SUCCESSION 3000+** ?

Base not exposed

MY Millions of years before present   S/G Subgroup   SST Sandstone

E

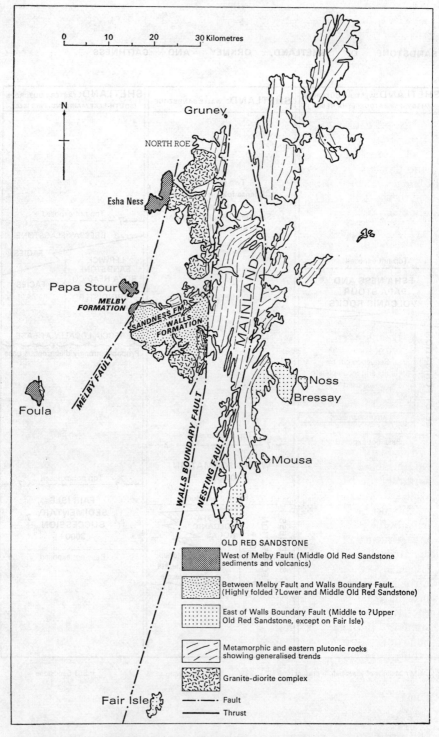

FIG. 10. *Geological sketch-map showing the outcrops of Old Red Sandstone rocks in Shetland and their structural relationships*

suggests that the rhyolites originated as flows or domes of obsidian. The rhyolitic tuffs are probably mainly lithified air-fall deposits, originally composed largely of pumice clasts.

Esha Ness

The Old Red Sandstone rocks of Esha Ness (Fig. 11) occupy the ground west of the probable northward extension of the Melby Fault, which here extends from Brae Wick north-north-eastwards to the mouth of Ronas Voe. The rocks consist mainly of a series of lavas, tuffs, agglomerates, and ignimbrites, and they are folded into a shallow north-north-east trending syncline, which plunges north-east in the northern part of the area and south-west in the southern. The sequence of lavas and tuffs is most fully developed in the southern part of the area, where the succession is as follows:

9. Andesitic lavas
8. Bedded andesitic agglomerate, tuff and subordinate sandstone
7. Andesite of The Bruddans and Stenness
6. Coarse andesitic tuff and bedded agglomerate
5. Andesite—thick on east limb, but thin and discontinuous on west limb of syncline
4. Mugearite—thick in north-west of Eshaness Peninsula, but thinning out eastwards. Absent on east limb of syncline
3. Ignimbrite and ? rhyolite. Exposed at the Grind of the Navir on the west limb of the syncline; intercalated with acid andesites (item 5) along the east limb of syncline
2. Intercalated olivine-basalts and andesites, with lenticular beds of tuff and tuffaceous sandstone (exposed on eastern limb of syncline only)
1. Reddish purple micaceous sandstones and tuffaceous sandstones.

The andesites are generally fairly acid and commonly decomposed. Both the tuffs and andesites give rise to spectacular cliffs and off-shore stacks. They are strongly jointed and break readily into large blocks which have been piled up on high-level storm beaches along the coast. The ignimbrite forming the Grind of the Navir contains flattened and welded pumice clasts and shards, which are well seen on the weathered surfaces (Plate X b).

The three lowest divisions of the Eshaness succession could be the equivalents of the Papa Stour rocks, but correlation over a distance of nearly 15 km in a sequence of lenticular lavas and pyroclastics must be regarded as highly tentative. It has been suggested that the volcanic rocks of Papa Stour and Esha Ness could be the time equivalents of the tuffaceous beds in the Upper Stromness Flags of Hoy in Orkney (Wilson and others 1935, p. 69), but the present writer believes that they are more likely to be equated with the volcanic rocks in the Eday Flagstones, particularly as pebbles of basic and acid lavas similar to those of West Shetland are common in the Middle and Upper Eday Sandstones of Eday (p. 87), which indicates that basic and acid lavas were exposed to erosion when the higher Eday sandstones were being laid down.

Foula

The greater part of the Island of Foula is formed of up to 1800 m of soft, buff-weathering sandstones with subordinate shales and siltstones (Fig. 12).

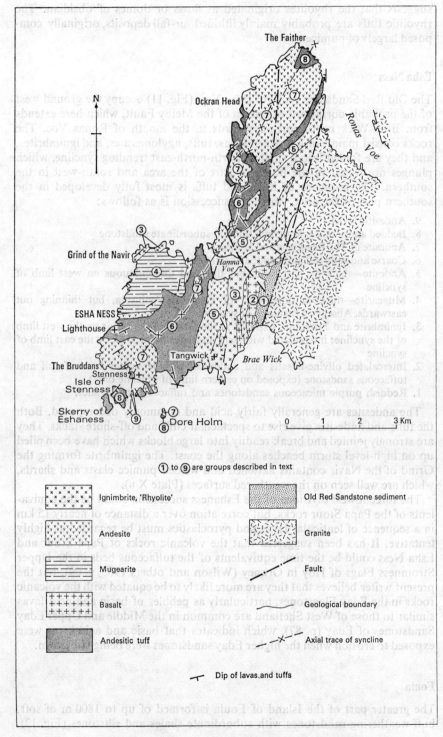

The Faither

Ockran Head

Ronas Voe

N

Grind of the Navir

Hamna Voe

ESHA NESS

Lighthouse

Tangwick

Brae Wick

The Bruddans

Stenness

Isle of
Stenness

Skerry of
Eshaness

Dore Holm

0　　1　　2　　3 Km

① to ⑨ are groups described in text

× × × × × × × × × × ×	Ignimbrite, 'Rhyolite'	Old Red Sandstone sediment
	Andesite	Granite
	Mugearite	Fault
+ + + + + + + + + + + + + + + + +	Basalt	Geological boundary
	Andesitic tuff	Axial trace of syncline

Dip of lavas and tuffs

FIG. 11. *Geological map of the Esha Ness District*

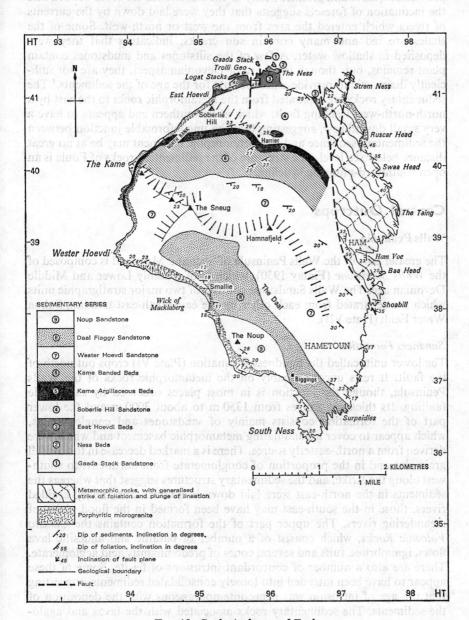

The legend within the figure reads:

SEDIMENTARY SERIES

- ⑨ Noup Sandstone
- ⑧ Daal Flaggy Sandstone
- ⑦ Wester Hoevdi Sandstone
- ⑥ Kame Banded Beds
- ⑤ Kame Argillaceous Beds
- ④ Soberlie Hill Sandstone
- ③ East Hoevdi Beds
- ② Ness Beds
- ① Gaada Stack Sandstone

Metamorphic rocks, with generalised strike of foliation and plunge of lineation

Porphyritic microgranite

↗20 Dip of sediments, inclination in degrees.
↙55 Dip of foliation, inclination in degrees
↘45 Inclination of fault plane
——— Geological boundary
—·—·— Fault

2 KILOMETRES
1 MILE

FIG. 12. *Geological map of Foula*

The sandstones are generally medium-grained and contain small scattered pebbles of quartz and granite. Many of the sandstones are cross-bedded and the inclination of foresets suggests that they were laid down by the currents of rivers which entered the area from the west or north-west. Some of the shales are red and many contain sun cracks, indicating that they were deposited in shallow water. A few of the siltstones and mudstones contain plant remains, but, though these are of Devonian aspect, they are not sufficiently diagnostic to provide a precise date for the age of the sediments.[1] The sedimentary rocks are separated from the metamorphic rocks to the east by a north-north-west trending fault, which at its southern end appears to have a very small downthrow, suggesting that the unconformable junction between the sedimentary sequence and the metamorphic basement may be at no great distance below sea level. The structure of the sedimentary rocks of Foula is an open southward-plunging syncline.

Central Outcrops

Walls Peninsula

The greater part of the Walls Peninsula of Shetland Mainland is composed of the *Walls Sandstone* (Finlay 1930), which is of possibly Lower and Middle Devonian age. The Walls Sandstone consists of two major stratigraphic units which are separated from each other by the east-north-east trending Sulma Water Fault (Plate VII).

Sandness Formation

The lower unit, called the Sandness Formation (Plate VI) crops out north of the fault. It rests unconformably on the metamorphic rocks of the Walls Peninsula, though the junction is in most places obscured by small-scale faulting. Its thickness ranges from 1350 m to about 3000 m and the lower part of the formation consists mainly of sandstones and conglomerates, which appear to cover an undulating metamorphic basement and which were derived from a north-easterly source. There is a marked decrease in the overall grain-size and in the proportion of conglomerate from north-east to south-west along the strike, and the sedimentary structures suggest that whereas the sediments in the north-east were laid down in alluvial fans and by braided rivers, those in the south-east may have been formed in the flood plains of meandering rivers. The upper part of the formation contains the *Clousta Volcanic Rocks*, which consist of a number of basaltic and andesitic lava flows, ignimbrites, tuffs and several cones of predominantly acid agglomerate. There are also a number of concordant intrusions of felsite. Some of these appear to have been intruded into loosely consolidated sediments, suggesting that the age of intrusion was penecontemporaneous with the deposition of the sediments. The sedimentary rocks associated with the lavas and agglomerates include some thick beds of mudstone and calcareous siltstone which may have been laid down in lakes ponded up by the newly erupted volcanic rocks.

---

[1] Recently identified spores from the Foula sediments suggest that these rocks are of Middle Old Red Sandstone age (R. N. Donovan, personal communication).

Walls Formation

The upper unit of the Walls Sandstone, termed the Walls Formation, crops out south of the Sulma Water Fault. It consists of up to 9000 m of highly folded, generally dark grey, fine-grained evenly bedded sandstone, which is in many instances interbedded with siltstone, shale and, locally, limestone. In thin section the sediments have some of the characteristics of greywackes, such as a high proportion of clastic matrix, and on a larger scale they have certain features in common with flysch deposits, though there are also a number of significant differences. The mode of deposition of these beds is not yet certain; the author believes that they may have been laid down in a fairly deep and extensive lake.

Poorly preserved fish remains determined by Dr R. S. Miles as acanthodian fish aff. *Cheiracanthus sp.* and crossopterigian and dipnoan scales have been found at a number of localities in the Walls Formation, and poorly preserved plant remains, principally of Hostimellid type occur in both formations of the Walls Sandstone. The fish remains suggest that the Walls Formation may be of Middle Devonian age, whereas the plant remains from the Sandness Formation, which include *Psilophyton sp.*, do not rule out a Lower Devonian age for the latter.

Structure

The rocks of the Sandness and Walls formations have been involved in two phases of intense folding. The first phase produced tight folds with east-north-east trending axes, as well as some east-north-east trending movement planes. The major structures ascribed to this period are the complex Walls Syncline and the Watsness–Browland Anticline (Plate VII). Within a 3 km-wide zone on either side of the Walls Syncline the fine-grained sediments have taken on a slaty cleavage and a lineation with an orientation which reflects the geometry of the major folds. Some areas are also affected by complex minor folding. The second period of deformation produced north-north-east to north trending folds, and in the area extending from Gruting Voe to Brindister Voe minor folds and cleavages developed during this period have deformed the earlier structures. In the more intensely folded areas the sediments have a mineral assemblage characteristic of the zeolite facies and, locally, the greenschist facies of regional metamorphism.

The rocks of the Walls Formation have been intruded by the Sandsting Complex (Fig. 9) which contains several large and many smaller enclaves of hornfelsed sediments. Thermal metamorphism has given rise to mineral assemblages characteristic of the hornblende-hornfels facies near the contact and alteration of a lower grade extends in places as far as 1·5 km from the granite. Both periods of folding in the Walls Sandstone are later than the emplacement of the complex which is dated at 360 to 370 million years. The deformation most probably took place during Upper Devonian time and is thus the latest phase of intense localised late-Caledonian folding recorded within the British Isles (see p. 69).

North Mainland

A number of small outliers of folded and faulted, predominantly sandy sediments crop out along and close to the north-east coast of North Roe

(Fig. 10). A rather larger outcrop forms most of the island of Gruney, 1·5 km N of the Point of Fethaland. This outcrop consists of a basal breccia overlain by conglomerate and arkose with thin beds of mudstone. These rocks may have been laid down first as scree, then as alluvial fan deposits and finally as the channel and overbank deposits of swift-flowing rivers. They are to some extent comparable with some of the sediments of south-east Shetland, but if Pringle's (1970, p. 166) contention that they are older than the Ronas Hill Granite is correct, they may be of roughly the same age as the Walls Sandstone.

Eastern Outcrops

South-east Mainland

The Old Red Sandstone of south-east Mainland forms a narrow discontinuous outcrop which extends from Rova Head, 3·5 km N of Lerwick, southwards for nearly 40 km to Sumburgh Head (Fig. 13). It also forms the islands of Bressay, Noss and Mousa. Fish remains, plants and '*Estheria*' have been obtained from over 50 different localities throughout the outcrop and it is believed that the age of the beds ranges from Middle Old Red Sandstone (i.e. Eday Beds age) upwards possibly into the Upper Old Red Sandstone.

It has long been the custom to divide these beds into the following sub-divisions:

(1) Basement Breccia, (2) Brindister Flags, (3) Rova Head Conglomerate, (4) Lerwick Sandstone, (5) Bressay Flags.

Although some of these divisions correspond to well-marked lithological types, others, such as the Brindister Flags, include a number of diverse rock types. The groups do not constitute a stratigraphic succession, and as is shown diagramatically in Fig. 14, the base of the sequence is formed by the Rova Head Conglomerate in the Lerwick area, by breccia in the Quarff–Fladda-bister district, and by the Brindister Flags, locally underlain by breccia, over most of the area south of Cunningsburgh. The Old Red Sandstone of south-east Shetland can be best visualised as a number of lithological facies of limited lateral extent which interdigitate with each other. They were laid down on an undulating basement of metamorphic rock, which at the time of deposition appears to have had an overall slope to the east-south-east. A north–south section at that time would have passed across an eastward-sloping scree-covered hill-shoulder in the Quarff district. This separated a deep south-eastward-flowing river valley in the north from a more open river valley, and later alluvial plain in the south. Low ground at times covered by a lake must have lain to the south-east of the present area and this lake periodically transgressed north-westward to form the relatively thin beds of fish-bearing calcareous flags which are intercalated with the terrestrial sediments.

Area south of Fladdabister. In the area south of Fladdabister (Fig. 13) the junction between the Old Red Sandstone and the metamorphic rocks is in part a fault and in part an unconformity. A basal breccia with angular pebbles of locally derived metamorphic and igneous rocks is developed in the Spiggie area, but elsewhere, as at Little Holm [380 097] in the extreme south,

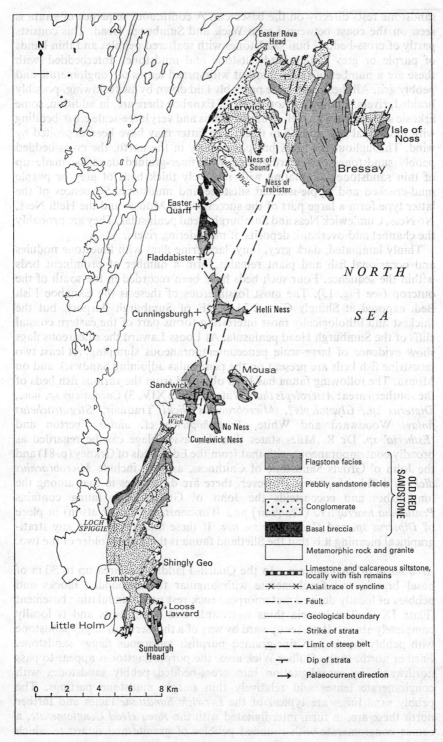

FIG. 13. *Geological map of the Old Red Sandstone of south-east Shetland (with palaeocurrent directions)*

sandstone rests directly on the basement. A continuous sequence of strata is seen on the coast between Leven Wick and Sumburgh Head. This consists partly of cross-bedded buff sandstones with scattered pebbles and thin bands of purple or grey sun-cracked siltstone and mudstone. Interbedded with these are a number of fairly thin but widespread lenses of conglomerate and pebbly grit. All these beds were probably laid down by swift-flowing, possibly braided, rivers. In the area north-east of Exnaboe there are, in addition, some arkosic sandstones with well-rounded grains and very large-scale cross-bedding with individual sets up to 5 m thick. The latter may have been deposited by wind. Throughout the area, but particularly in the north, the cross-bedded pebbly sandstones are interdigitated with finer-grained deposits, made up of thin sandstones alternating with relatively thick beds of grey or purple mud-cracked and ripple-marked siltstone and mudstone. Sequences of the latter type form a large part of the succession of Mousa, and the Helli Ness, No Ness, Cumlewick Ness and Sumburgh Head peninsulas. They are probably the channel and overbank deposits of meandering rivers.

Thinly laminated, dark grey, limy, lacustrine flags with limestone nodules and occasional fish and plant remains form a number of prominent beds within the sequence. Four such beds have been recorded in the south of the outcrop (see Fig. 13). The most fossiliferous of these is the Exnaboe Fish Bed, exposed at Shingly Geo, 2·5 km NE of Sumburgh Airport, but the thickest and lithologically most interesting forms part of the eastern coastal cliffs of the Sumburgh Head peninsula. At Looss Laward the calcareous flags show evidence of large-scale penecontemporaneous slumping. At least two lacustrine fish beds are present on the peninsulas adjoining Sandwick and on Mousa. The following fauna has been obtained from the various fish beds of the southern area: *Asterolepis thule* Watson (Plate XIV, 3) *Coccosteus sp. nov.*, *Dipterus sp.*, *Glyptolepis?*, *Microbrachius dicki* Traquair, *Stegotrachelus finlayi* Woodward and White, *Tristichopterus* cf. *alatus* Egerton and '*Estheria*' *sp.* Dr R. Miles states that this assemblage can be regarded as broadly contemporaneous with that from the Eday Beds of Orkney (p. 81) and the John o' Groats Sandstone of Caithness, as both include *Microbrachius dicki* and *Tristichopterus*. However, there are differences in that among the lung fishes and coccosteids the John o' Groats–Eday fauna contains *Pentlandia macroptera* (Traquair) and *Watsonosteus fletti* (Watson) in place of *Dipterus sp.* and *Coccosteus sp. nov.* If these differences have any stratigraphical meaning it is that the Shetland fauna is the slightly older of the two.

Area north of Fladdabister. In the Quarff–Fladdabister area up to 30 m of basal breccia and fanglomerate with angular to subrounded blocks and pebbles of locally derived metamorphic rock rest on an undulating basement (Plate IX A). The breccia thins westwards and southwards and is locally completely absent. It passes upward by way of a thin series of flaggy sandstone with pebbly bands into fine-grained purplish micaceous flaggy sandstone. Farther north, in the Gulber Wick area, the purple flagstones appear to pass northwards by interdigitation into cross-bedded pebbly sandstones with conglomerate lenses and relatively thin purple mudstone partings. The pebbly sandstones are typical of the *Lerwick Sandstone* facies and farther north these are, in turn, interdigitated with the *Rova Head Conglomerate*, a coarse conglomerate with rounded pebbles of granite and quartzite which

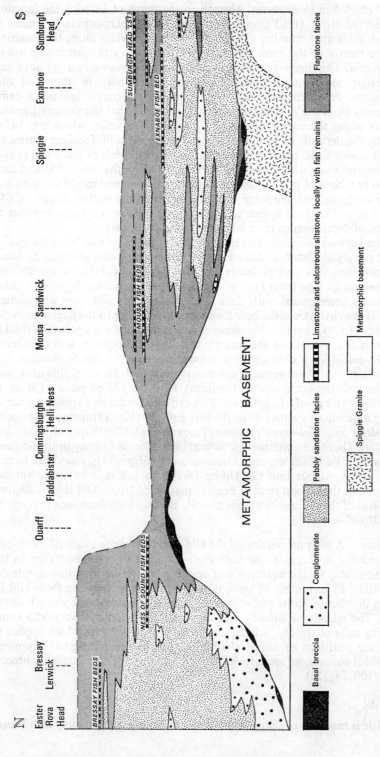

FIG. 14. *Hypothetical section showing the possible facies relationships within the east Shetland Old Red Sandstone*

Flagstone facies

Limestone and calcareous siltstone, locally with fish remains

Pebbly sandstone facies

Metamorphic basement

Spiggie Granite

Conglomerate

Basal breccia

METAMORPHIC BASEMENT

S

Sumburgh Head

Exnaboe

Spiggie

Sandwick

Mousa

Cunningsburgh
Helli Ness

Fladdabister

Quarff

Bressay
Lerwick

Easter
Rova Head

N

SUMBURGH HEAD TST

EXNABOE FISH BED

MOUSA FISH BEDS

NESS OF SOUND FISH BEDS

BRESSAY FISH BEDS

locally reach 1 m in diameter. Though to the north of Lerwick the junction between the Rova Head Conglomerate and the metamorphic basement is sheared, there are a number of exposures of basal breccia along this junction and the matrix of the most westerly exposures of conglomerate consists of fine breccia. This suggests that the conglomerate passes down, without any intervening sediment, into basal breccia, which may be thin and discontinuous. Palaeocurrent data suggest that these coarse sediments came principally from the north-west, and it is probable that the present Lerwick area lay along the course of a major intermontane valley which was being filled by the deposits of torrential rivers. On the Ness of Trebister sediments of the Lerwick Sandstone facies are interbedded with beds of black and purple siltstone and shale with abundant '*Estheria*', and on the Ness of Sound they contain two beds of nodular limy lacustrine siltstone and mudstone with fish remains belonging to the same species as those in the southern area (p. 64). These fine-grained beds appear to die out north-westwards, suggesting that at the time of deposition an open basin lay to the south-east.

In the Lerwick peninsula and on the islands of Bressay and Noss, the coarse fluvial pebbly sandstones of the Lerwick Sandstone facies give way by inter-digitation upwards and probably also southwards to beds of the *Bressay Flagstone* facies. The latter consist of generally flaggy but locally cross-bedded sandstones interbedded with dark grey and purple siltstones and shales. Some of the rhythmic units bear a certain resemblance to the flagstone cycles of Orkney (p. 74), but here the channel sandstone phases are generally thicker and with few exceptions the fine-grained beds are coarser, less calcareous and devoid of such characteristic minor structures as fine lamination and syneresis cracks. Plant remains are common in both facies of this area, and these include the conspicuous 'Corduroy Plant', a ribbed plant stem up to 60 cm long and nearly 10 cm wide. Fish remains found in a small number of closely adjoining localities in north-east Bressay (Fig. 13) include the forms *Asterolepis sp.*, *Holonema ornatum* Traquair and *Glyptolepis* cf. *paucidens* (Agassiz). There is no comparable assemblage either in Orkney or the Scottish mainland. Two of the genera, *Asterolepis* and *Glyptolepis*, elsewhere have a range which extends into the Upper Devonian and it is possible, but not proved, that the highest beds of Bressay may be of Upper Old Red Sandstone age. Most of the sediments ascribed to the Bressay Flagstone facies appear to be of fluvial origin.

Structure.   A high proportion of the Old Red Sandstone strata of south-east Shetland dip to the east and east-south-east, with the steepest dips in the extreme south. In the northern and central parts of the area this simple dip is modified by a number of open north to north-east trending folds and by a large number of faults and shatter belts with a predominantly north–south trend. The islands of Bressay and Noss are traversed by two north–south trending belts of steeply inclined and locally inverted strata which appear to mark the positions of major shatter belts in the metamorphic basement, and which contain, or are associated with, extensive areas of tuffisitic breccia (see p. 100, Fig. 26).

Fair Isle

Fair Isle is made up of sedimentary rocks of Middle and possibly also Lower

West coast of Walls Peninsula, Shetland, between Mu Ness and Bay of Deepdale. Steeply inclined sandstones of Sandness Formation (D964)

Plate VIII

Geo of Bordie, north-west coast of Papa Stour, Shetland. Bedded rhyolitic agglomerate resting on weathered eroded surface of rhyolite (D926)

A. Basal Middle Old Red Sandstone breccia resting on eroded schists of Quarff Nappe Succession, Fladdabister, east coast of south Mainland, Shetland (D1325)

Plate IX

B. Pseudonodules in Middle Old Red Sandstone sandstones and siltstones. East shore of West Voe of Sumburgh, south Mainland, Shetland (D1307)

Old Red Sandstone age, which are for the most part steeply inclined to the east-south-east. At least 2700 m of strata are present and these can be divided into four lithostratigraphical units (Fig. 15). All four groups contain a high proportion of grey to buff, locally red-stained, arkosic sandstone and along the north-west coast of the island as well as in Bu Ness and Vaasetter in the east there are a number of beds of pebbly grit and conglomerate. The two lower groups contain beds of dolomitic mudstone and siltstone which are relatively thin and widely spaced in the lowest part of the sequence, but form a high proportion of the Observatory Group in which bands of predominantly fine-grained sediment are up to 200 m thick. In the northern part of the Bu Ness Peninsula the sequence has a strongly rhythmic character with beds of sandstone, up to 9 m thick, alternating with much thinner beds of siltstone and mudstone. The fine-grained rocks of the Observatory Group have yielded remains of the plants *Dawsonites roskiliensis* Chaloner, *Hostimella sp.* and *Thursophyton milleri* (Salter); the branchiopod crustacean '*Estheria*'; and unidentifiable cycloid fish scales. The Bu Ness Group has yielded *Hostimella sp.*, cf. *Prototaxites sp.*, *Svalbardia scotica* Chaloner, *Zosterophyllum*? and '*Estheria*', as well as scales of dipnoan fish and a plate of a coccosteid arthrodire. The flora from the Observatory Group could be of Lower Devonian age, but both the fish and plants from the Bu Ness Group have a Middle Devonian aspect.

Palaeocurrent indicators suggest that the sediments were deposited by currents moving in an easterly to north-easterly direction. The pebbly sandstones and conglomerates were probably deposited by braided rivers in alluvial fans close to the western margin of a lake basin. The fine-grained dolomitic beds were probably laid down during periods when the lake waters encroached westwards across the distal edges of these fans. The rhythmic sequences of Bu Ness may be the deposits of a river or rivers meandering over an alluvial plain.

Over the greater part of Fair Isle the strata are steeply inclined to the east-south-east, but in the south-west corner of the island they are flexured into open, eastward-plunging folds. In this area the fine-grained sediments have a slaty cleavage and a lineation, the geometry of which reflects that of the major folds. Fair Isle is cut by a number of near-vertical west-north-west trending faults, along which the deep geos on the west coast of the island have been excavated. Most of these faults do not have large throws, but one may have a dextral transcurrent displacement of 550 m. Some of the fault planes have basic or acid dykes emplaced along them and many contain, or are associated with, a network of scapolite-carbonate veins. The sedimentary rocks cropping out in the south-west corner of the island are indurated either as a result of the tectonic deformation of the area or by the thermal metamorphism produced by a granitic mass which may crop out beneath the sea just south-west of Fair Isle.

Correlation

The three groups of Old Red Sandstone rocks of Shetland Mainland differ from each other in age, in their depositional and volcanological development, in their tectonic history and in the extent to which they have been affected by igneous intrusions. As all three appear to rest directly on the metamorphic

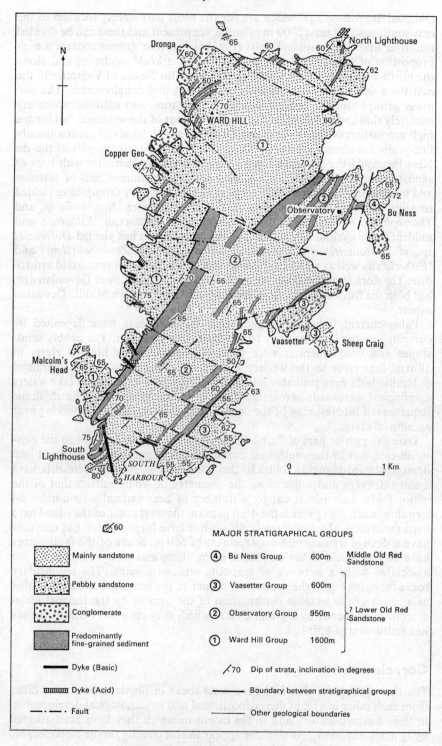

FIG. 15. *Geological map of Fair Isle*

basement, they do not represent parts of a continuous sequence but form the deposits of three geographically distinct basins. The rocks of Melby and Foula appear to have been laid down in the most southerly of these; they may in fact have been deposited along the north-western fringe of the extensive shallow and tectonically stable Orkney–Caithness basin. The Walls Sandstone and possibly also the Fair Isle sandstone may have been deposited farther north in an intermontane, tectonically and volcanologically active basin, which developed somewhat earlier than the others and was affected towards or just after the end of Middle Old Red Sandstone times by two phases of compression. The East Shetland sediments were probably laid down on the western margin of the most northerly basin which was bounded by mountainous terrain to the west, but had access to an open lake to the south-east. The present virtual juxtaposition of three such diverse groups of rocks is most readily explained by postulating subsequent dextral transcurrent movements along both the Melby and Walls Boundary faults. The extent of movement along the latter may have been in the order 60 to 80 km.

REFERENCES

Chaloner 1972; Finlay 1926b, 1930; Fleming 1811; Flett 1908; Flinn 1961a; Flinn and others 1968; Geikie 1879, 1897; Miles and Westoll 1963; Murchison 1853, 1859; Mykura 1972b; Mykura and Phemister 1976; Peach and Horne 1879a, 1884, *in* Tudor 1883; Peach, C. W. 1879; Pringle 1970; Traquair 1908; Watson 1934; Westoll 1937, 1951.

8. OLD RED SANDSTONE OF ORKNEY[1]

Probable Lower Old Red Sandstone

At Stromness and Graemsay (Fig. 16) the crystalline basement rocks which form the hilly pre-Old Red Sandstone land surface of Orkney (p. 40) are overlain by thin layers and lenses of coarse breccia, conglomerate and sandstone. These pass laterally into sandy flagstones of Stromness type. At Yesnaby, however, only the upper parts of the original hills are covered by an apron of breccia which is of the same age as the Stromness Flags. The lower parts of these hills are overlain by a series of older sandstones and subordinate conglomerates, which Fannin has termed the *Yesnaby Sandstone Group*, and which are themselves separated by an angular unconformity from the basal beds of the Stromness Flags.

The Yesnaby Sandstone Group contains two facies which are now separated from each other by the Garthna Geo Fault (Fig. 16). The beds exposed to the south of the fault are termed the *Harra Ebb Formation*. They rest on an irregular surface of the crystalline basement which forms part of the steep western slope of an exhumed hill and, farther west, part of the flattish plain at the foot of that hill. The beds are composed of up to 100 m of interbedded sandstones and siltstones with tongues and lenses of breccia and conglomerate near the base. Palaeocurrent data indicate that the beds were laid down by currents travelling directly down the hillside. It is thus possible that they originated as alluvial fan and talus deposits around the base of the hill. They are unconformably overlain by the basal sandstones and pebbly sandstones of the Stromness Flags (p. 73) and there is an angular discordance of 6° to 10° between the two formations.[2] The rocks north of the Garthna Geo Fault have been termed the *Yesnaby Sandstone Formation*. They comprise two sandstone units with distinctive characters. The lower unit consists of rusty-weathering grey fine- to medium-grained well-graded sandstone with large-scale, predominantly tabular, cross-bedding. The foresets are steeply inclined and individual sets range in thickness from 1 to 3 m. At intervals of 5 to 6 m the cross-bedding is truncated by major bedding plane surfaces, which, it is believed, were originally horizontal but have since been tilted to dip at 14°

---

[1] The sections dealing with the probable Lower Old Red Sandstone and the Stromness Flags are based to a large extent on the work of Dr N. G. T. Fannin (Fannin 1970) and the section on the Eday Beds incorporates data supplied by Miss J. M. Ridgway. In consequence the classification used in this account differs in some respects from that published by Wilson and others (1935). Information provided by Mr U. McL. Michie of the Institute's Geochemical Division has been used throughout the chapter.

[2] Recent mapping by Michie and Mr. N. L. Watts (personal communications) suggests that most of the sediments overlying the Basement hills south-east of Garthna Geo belong to the Yesnaby Sandstone Group. Michie believes that most of these rocks are older than the Harra Ebb Formation.

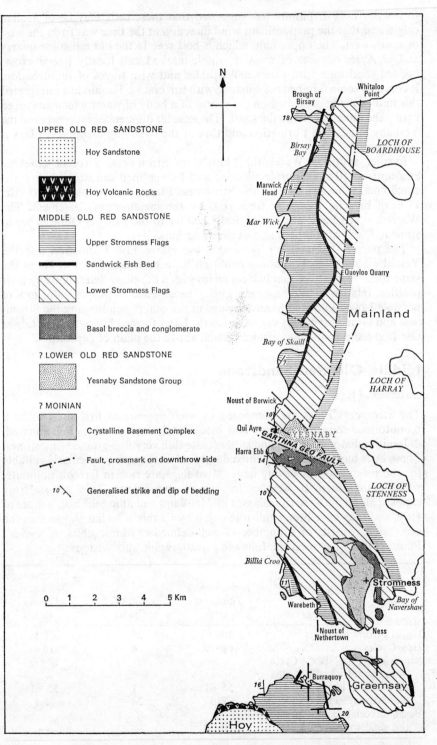

N

UPPER OLD RED SANDSTONE

Hoy Sandstone

Hoy Volcanic Rocks

MIDDLE OLD RED SANDSTONE

Upper Stromness Flags

Sandwick Fish Bed

Lower Stromness Flags

Basal breccia and conglomerate

? LOWER OLD RED SANDSTONE

Yesnaby Sandstone Group

? MOINIAN

Crystalline Basement Complex

Fault, crossmark on downthrow side

Generalised strike and dip of bedding

0 1 2 3 4 5 Km

Whitaloo Point
Brough of Birsay
18
Birsay Bay
LOCH OF BOARDHOUSE
Marwick Head
6
Mar Wick
8
Quoyloo Quarry
Mainland
Bay of Skaill
7
LOCH OF HARRAY
Noust of Borwick
10
Qui Ayre
YESNABY
GARTHNA GEO FAULT
Harra Ebb
14
10
LOCH OF STENNESS
Billia Croo
11
Stromness
Bay of Navershaw
Warebeth
Ness
Noust of Nethertown
Burraquoy
16
Graemsay
20
Hoy

FIG. 16. *Geological map of the western seaboard of Orkney Mainland and northern Hoy*

to the north-west. Fannin has suggested that these beds may be of aeolian origin and that the predominant wind direction at the time was from the west or north-west. The upper unit, which is best seen in the old millstone quarry at Qui Ayre, consists of massive, ripple-marked and locally trough-cross-bedded sandstone with rare small pebbles and with traces of bioturbation. It contains some thin beds of siltstone with sun cracks. Fannin has interpreted this unit as being deposited on the shores of a body of water which advanced across the dune-field from the south. The angular difference in dip between the Yesnaby Sandstone Formation and that of the overlying Stromness Flags is 10 degrees.

Further south at Warebeth, 2 km W of Stromness, a recent borehole encountered 50 m of purple siltstone and fine-grained sandstone below the conglomerate at the base of the Stromness Flags. The purple beds overlie 11 m of breccia, which, in turn, rests on the metamorphic basement. The Warebeth 'red beds' are lithologically similar to the ? Lower Old Red Sandstone red beds of southern and western Caithness.

The purple sediments of Warebeth, the Harra Ebb Formation and the Yesnaby Sandstone Formation could all have been laid down at about the same time. They are probably local facies with a lithology determined by their position relative to the basement hills. The absence of purple sediments of possible Lower Old Red Sandstone age in the outcrops adjoining the Stromness and Graemsay inliers suggests that the latter formed hills which, in Lower Old Red Sandstone times, projected well above the plain of deposition.

Middle Old Red Sandstone

Stromness Flags

The Stromness Flags are composed of what appears at first sight to be a monotonous sequence of grey and black thinly bedded, in part laminated, dolomitic siltstones, shales and subordinate thin very fine-grained sandstones. These beds have been aptly named flagstones as they are particularly suitable for the production of paving flags. Most flags are rich in ferroan dolomite, which causes them to weather to an ochreous colour. They are superbly exposed along the western seaboard of Mainland and along the north shore of Hoy, and they have been studied in detail by Fannin who has shown that the succession consists of a number of well-defined rhythmic units or 'cycles'. Fannin (1970) has used the following stratigraphic subdivisions:

	Approximate Thickness (m)	No. of Cycles	Average Thickness of cycles (m)
Upper Stromness Flags	190+	20+	8·5
Hoy Cycles	61–79	4	16·6
Sandwick Fish Bed Cycle (with Sandwick Fish Bed 2·5 m above base)	55–61	1	55–61
Lower Stromness Flags	215	24	8·1
Basal breccia, conglomerate and sandstone	0–20	—	—

A. Dore Holm, Stenness, off north-west coast of Shetland Mainland. Off-shore holm with natural arch, composed of andesitic agglomerate capped by andesite (D1656)

Plate X

B. Ignimbrite (welded rhyolitic tuff). Grind of the Navir, north-west coast of Shetland Mainland (D1662)

The Lower Stromness Flags have been taken to be of Upper Eifelian age and the beds above the Sandwick Fish Bed have been classed as Lower Givetian (Table II; Westoll 1951; Miles and Westoll 1963). Fish remains are present throughout the sequence and they are particularly abundant and well preserved in the Sandwick Fish Bed. The fauna recorded from the various subdivisions is as follows:

Upper Stromness Flags: *Coccosteus?*, *Dickosteus threiplandi* Miles and Westoll, *Dipterus valenciennesi*, *Glyptolepis paucidens* and *Homostius milleri*.

Hoy Cycles: *Dickosteus threiplandi*, *Dipterus sp.* and *Glyptolepis sp.*

Sandwick Fish Bed Cycle: *Cheiracanthus murchisoni* Agassiz, *Cheirolepis trailli* Agassiz, *Coccosteus cuspidatus*, *Diplacanthus striatus* Agassiz, *Dipterus valenciennesi*, *Gyroptychius agassizi*, *G. microlepidotus* (Agassiz), *Osteolepis macrolepidotus* Agassiz (Plate XIII 3), *Pterichthyodes milleri* (Agassiz) and *Rhadinacanthus sp.*

Lower Stromness Flags: *Coccosteus cuspidatus*, and *Dipterus sp.*

Large plant fragments are relatively rare and poorly preserved and the principal genera recognised are *Hostimella*, *Thursophyton* and *Protopteridium*. These cannot be used for accurate dating. Spores are abundant in many of the siltstones and Richardson (1965) and Fannin (1970) have recorded a large number of species. Apart from giving general support to the dating by the fossil fish, they have not so far provided any evidence for the exact age of the Stromness Flags, but may in future form the basis for a more precise stratigraphic zonation.

Basal Breccias, Conglomerates and Sandstones

The basal beds of the Lower Stromness Flags vary greatly both in thickness and in the size of their clasts. They are well exposed on the north-western margin of the Graemsay inlier and along the shore at Ness, at the south end of the Stromness inlier (Fig. 16). In the Yesnaby area a thin layer of basal breccia rests directly on the basement rocks which form the local hills, but along the coast thin beds of pebbly sandstone at the base of the Stromness Flags rest on the sediments of the Yesnaby Sandstone Group.

At Ness and north Graemsay the basal breccia ranges from 5 to 20 m in thickness and locally fills hollows in the old land surface. The breccia passes laterally and upward into conglomerate with small pebbles, which is inter-fingered with pebbly sandstones and thin lenses of sun-cracked siltstone. The breccias and conglomerates are composed of unsorted angular to subrounded boulders and cobbles of locally derived granite, granite-gneiss and, more rarely, schist, set in a matrix of arkose which in places is composed entirely of fine granite debris. At Graemsay some clasts are coated with a thin layer of interlaminated dolomite and siltstone which in places has a mammillated structure. These coatings appear to be stromatolites (Fannin 1969, p. 82). The breccias with their angular clasts are interpreted as scree and talus deposits which accumulated at the flanks of the basement hills and were later partly re-worked and rounded on the shore of the lake surrounding these hills. The stromatolite coatings of the clasts were formed in shallow water and the complete coating of pebbles suggests that the latter were being constantly rolled by waves.

Flagstone Facies

Above the basal breccias the Stromness Flags are made up of over 50 cycles which are thought to have been formed by the fluctuations in the level of a single large and generally shallow lake in the Orkney–Caithness area. The cycles are basically similar to those described in the Caithness Flagstones by Crampton and Carruthers (1914, pp. 89–93) and the lithological characters of such a cycle are shown in Fig. 17. The base of the cycle is taken at the base of the black thinly laminated fine sediment which commonly contains fish remains and was laid down during a quiescent period when little or no coarse sediment entered the lake and when the water, though still shallow, was at its deepest. The sequence of sediments above this suggests that thereafter the depth of the lake gradually decreased and that eventually an influx of coarser sediment brought in by a river or stream pushed a delta out over the shallow or dried-up lake floor. It is believed that the Orcadian Basin was a rapidly sinking, tectonically controlled basin, and that the rhythmic sedimentation was regulated by an interplay of tectonic and climatic changes.

The Stromness Flags contain a number of features which can be used for local correlation. The most obvious for detailed correlation are the 'deepest water' fish-bearing facies of the individual cycles which are extremely persistent laterally. Individual cycles can sometimes be recognised in neighbouring sections by their total thickness, which also remains constant over long distances. There are also a number of marker horizons which maintain their character for some distance laterally. Thus in the Lower Stromness Flags two thinly laminated iron-rich beds of silty dolomite, which are rusty orange-weathering and contain large chert nodules, occur at 14 and 59 m below the Sandwick Fish Bed. A third marker horizon, just above the base of the Sandwick Fish Bed, is a 25 cm-thick calcite mudstone, which weathers to a distinctive bluish grey colour. The fourth marker is a massive 2 m-thick bed of bluish grey to black silty mudstone, which occurs at 56 m above the Sandwick Fish Bed throughout the area. In addition, several thin bands of green tuff crop out in Hoy at horizons 3 and 170 m above the Sandwick Fish Bed. These, however, thin out northwards and extend only into the southern end of Mainland.

The thickness of individual cycles commonly ranges from 5 to 10 m but the Sandwick Fish Bed cycle has a thickness of 60 m. Each cycle can be divided into two major lithological facies. The lower of these was deposited under a continuous cover of water; the upper in shallow water which at times dried up completely.

The *lower facies* commences with (1) up to 1 m of dark grey to black silty mudstone interlaminated with siltstone or fine sandstone. It generally has graded laminae which may have either a high bitumen and pyrite or a high carbonate (usually ferroan dolomite) content. It contains either complete or fragmentary fish remains. This deposit was laid down in relatively quiet and sometimes stagnant waters on a lake bottom undisturbed by wave action. The lake waters may, at times, have been thermally stratified and some of the graded laminae may have been deposited by turbidity currents. These 'quiescent water' beds grade upwards into (2) thinly interbedded bituminous silty mudstones and fine sandstones together with some discrete beds of massive siltstone and fine sandstone, some of which fill small erosional channels. This part of the sequence is characterised by the presence of numerous small sub-aqueous shrinkage (syneresis) cracks (see Donovan and Foster 1972) which were infilled by sand or silt and were then compacted and contorted. Algal stromatolite sheets and mounds are common in these beds and scattered fish fragments are generally present. Sediments of this type were laid down in water in which bottom currents were spasmodically active. In some cycles these sediments are interbedded with massive beds of calcareous siltstone which are rich in spores and comminuted plant debris.

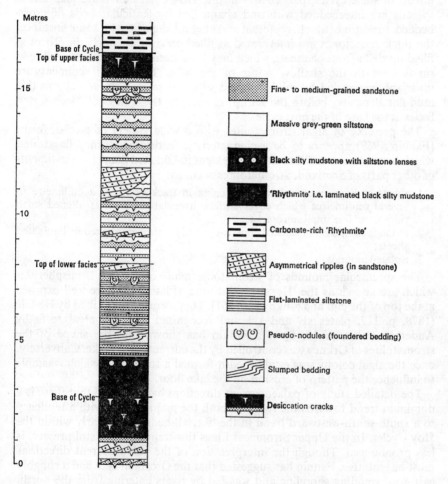

FIG. 17. *A measured cycle in the Stromness Flags*

The massive siltstones were probably formed in relatively quiet shallow water close to the lake shore which formed traps for the accumulation of 'vegetable hash'. None of the beds in the lower facies have sun cracks, which indicates that the water of the lake never receded completely.

In most cycles the *upper facies* commences with (1) thinly banded sediments which are lithologically very similar to the higher beds of the lower facies, but have, in addition, well developed sand-filled desiccation polygons

(Plate XII A). In many instances stromatolite sheets and mounds cover the cracked surfaces and extend down into the sun cracks. These beds were probably laid down in quiescent waters close to the lake margin, where the shallow water periodically receded to leave coastal mud flats. In most instances they are overlain by (2) ripple-cross-laminated sandstones and silt-stones with desiccation cracks. These beds appear to be alluvial flood-plain or delta-top deposits laid down by rivers which entered the lake from the north. In some cycles, particularly in the 'Hoy Cycles' of Hoy, such fluvial spreads are interbedded with and channelled by lenticular beds of cross-bedded sandstone which represent channel-fill deposits. In some instances the thick sandstones are interpreted as filled-up river channels, in others as filled-up delta-front channels, which may have been cut in the soft muds and sands beneath the shallow water of the lake. The fluvial sediments are invariably succeeded by thinly banded silts and sands with sun cracks (i.e. mud-flat deposits) before the abrupt change to the quiescent 'deep-water' facies at the base of the next cycle.

The presence of algal stromatolites with a wide variety of growth forms (Fannin 1969) appears to be a characteristic feature of Orkney flagstones, which has not been noted to the same extent in Old Red Sandstone sediments of other parts of Scotland. Stromatolites occur as:

1. sheets of limestone or dolomite ranging in thickness from a millimetre to several centimetres which cover surface irregularities such as aligned small ridges, and line small sun cracks;
2. mounds composed of stacked convex-upward hemispheres linked by laminated sheets;
3. isolated mounds.

The spectacular mounds of stromatolite made of stacked hemispheroids which are known as the 'Horse-tooth Stone' (Plate XII B) are well exposed at the top of the cliff at Yesnaby [220 161]. They were first described by Heddle (1878, p. 117, plates xiv and xv) and recognised as fossil algal reefs by Anderson (1950, p. 10, fig. 6). Fannin has shown that the shape of the stromatolites of Orkney was controlled by the currents in the lake. Conversely, once the algal colony was established it formed a tough mat which was able to influence the pattern of erosion on the lake floor.

The detailed study of palaeocurrent directions has indicated that there is a dominant trend towards the south in both the major facies, with a tendency to a south-south-westward trend in the 'fluvial' beds, particularly within the Hoy Cycles. In the Upper Stromness Flags this trend, though still present, is less pronounced. Though the interpretation of the palaeocurrent directions must be tentative, Fannin has suggested that the Orcadian lake had a roughly east–west trending shoreline and was fed by rivers entering from the north. The periodic retreat of the shoreline was probably due to the southward advance of the delta-fronts. Within the lake the dominant current flow was offshore and there is little evidence for any long-shore currents.

The 'typical' cycle described above is not universally found and there is considerable variation both in time and space. The simplest cycles occur in the Lower and Upper Stromness Flags. The Hoy Cycles are considerably thicker than either of the other two groups and they have a higher proportion of lower facies sediments. They contain channel sandstones which on Hoy reach a thickness of 11 m. This indicates that during the deposition of the

Hoy Cycles a large river sytem persisted over or near western Orkney. By far the thickest cycle in the Stromness Flags is that which contains the Sandwick Fish Bed. It ranges in thickness from 55 m in the south to 61 m in the north, though the fish bed *sensu stricto* is a finely laminated carbonate-rich band only 50 cm thick and located some 2·5 m above the base of the cycle. The cycle contains up to 55 m of true lacustrine (i.e. lower facies) sediments and, as it has been confidently correlated with the Achanarras and Niandt Fish Beds of Caithness and the Melby Fish Bed of Shetland (Miles and Westoll 1963), it must represent a period when the lake was considerably deeper and more extensive than at other times during the deposition of the Stromness Flags. The maximum water depth may have been 50 m.

In the past the Sandwick Fish Bed was worked in a large number of small quarries in West Mainland for paving flags and roofing slates. At present only the quarry at Quoyloo (Fig. 16) is still active. This still yields abundant fish remains, though modern methods of quarrying are less conducive to the preservation of good specimens. The fish bed can be recognised on the quarry face by its rusty weathering. Fish remains may be evenly scattered on some bedding planes, on others they may be so abundant as to form a matted mass. On joint faces the fish can be recognised as black coal-like lenses, sometimes up to 1 cm thick. Black bituminous matter is often present along the joint planes of the fish bed, and in the more compact carbonate-rich portions of the bed a little oil is also found.

Rousay Flags

The Rousay Flags (referred to in the Geological Survey memoir as the Rousay Beds) consist of over 1500 m of rhythmically bedded, predominantly fine-grained 'flagstones' which on lithological criteria alone are extremely difficult to distinguish from the underlying Stromness Flags. The 'Rousay Beds' were set up as a separate formation by Flett (1898b), who noted that the upper part of the Orkney flagstone succession has a less abundant fauna than the lower part and that this fauna contains the two fish species *Thursius pholidotus* Traquair and *Coccosteus* (now *Millerosteus*) *minor* (Miller) (Plate XIV, 4) as well as the branchiopod crustacean '*Estheria*' *membranacea* Pacht (now *Asmussia*). All these forms are unknown in the underlying Stromness Flags. Flett did not attempt a precise definition of the base of the Rousay Beds, but the Geological Survey (Wilson and others 1935, p. 18) have fixed a provisional line 'at the base of a band of limy flags rich in fragmentary fish and plant remains that lies immediately below the lowest bed in which '*Estheria*' *membranacea* has been found'. The presence of '*Estheria*' was, in fact, taken as the main criterion for including isolated outcrops in the Rousay Flags.

The Rousay Flags form the bedrock of about a third of the total area of the Orkney Islands (Fig. 3). They form the greater part of the northern islands except Eday, the eastern quarter of West Mainland and large portions of East Mainland, Burray, South Ronaldsay, Flotta and east Hoy. They are believed to be the stratigraphical equivalents of the Mey and Ham–Scarfskerry subgroups of the Caithness Middle Old Red Sandstone (Donovan and others 1974) and have been classed as Givetian. In addition to the diagnostic fossils mentioned above the Rousay Flags have yielded the following forms:

Asterolepis orcadensis Watson (Plate XIV, 5), *Cheirolepis sp.*, *Glyptolepis paucidens*, *Gyroptychius sp.*, *Homostius milleri*, *Mesacanthus peachi* Egerton, *Osteolepis panderi* Jarvik, and *Thursius sp.*

Though rare fish remains are to be found in the 'quiescent water' facies of most cycles in the Rousay Flags the Geological Survey noted a number of 'fish beds' which contain certain species in abundance. Thus in Rousay three bands with abundant remains of *Millerosteus minor* occur at horizons which are said to be about 360, 450 and 456 m above the base of the group (Fig. 18). Similar fish beds with *M. minor* occur in Westray and Papa Westray. In Eday, Stronsay and on the east coast of Deerness fish remains with abundant specimens of a small form of *Dipterus valenciennesi* and rare samples of *Asterolepis orcadensis* are present in the flags immediately below the top of the group.

Examples of '*Estheria*' are abundant in a number of the ochreous-weathering, fish-bearing 'quiescent water' facies beds within the lower part of the Rousay sequence, particularly in the north-east corner of West Mainland (between Evie and Wide Firth) and in southern Rousay. Scattered '*Estheria*' have been recorded on most islands and they occur throughout the sequence. In South Walls (Hoy) they are present in a bed thought to be only 60 m below the top of the group. Plant remains are fairly abundant but they appear to be even more fragmentary and more poorly preserved than in the Stromness Flags. Only fossil wood, *Hostimella* and *Thursophyton* have been recognised. Stromatolites have now been recorded throughout the Rousay Flags and they are particularly common on Sanday.

Lithology

The cyclic units of the Rousay Flags are closely similar to those of the Stromness Flags, but the following minor differences have been used to distinguish them:

1. The Rousay Flags commonly weather to a grey colour which contrasts with the predominant ochreous-weathering of the Stromness Flags and indicates that their carbonate content is mainly calcite, rather than ferroan dolomite.
2. The fish-bearing 'quiescent water' facies of the Rousay cycles weathers in places to a purplish colour and the fish beds are commonly impure limestones.
3. The effects of differential weathering of the hard and soft members of individual cycles are more pronounced in the Rousay Flags than in the Stromness Flags and terrace features are prominent on the hillsides of Westray, Rousay and West Mainland. Terraces are, however, much less obvious in the Rousay outcrops further east and south.

Though sandstone does not, as a rule, form an appreciably greater proportion of the cyles of the Rousay Flags than the Stromness Flags, thick and, in places pebbly, sandstones are present in the higher beds of the group on the island of Rousay (Fig. 18). On this island the 90 m of strata immediately below the highest exposed fish bed contain several bands of pebbly sandstone. The pebbles are up to 5 cm in diameter and are composed mainly of quartz, granite and schistose rocks. The pebbly sandstones thin out and decrease in grain size in a north-westerly direction and they do not reappear at the probable equivalent horizon in Westray. Other prominent sandstones can be seen at the north-east end of North Ronaldsay where they are fine- to medium-grained, reddish in colour, and can be traced laterally for 800 m.

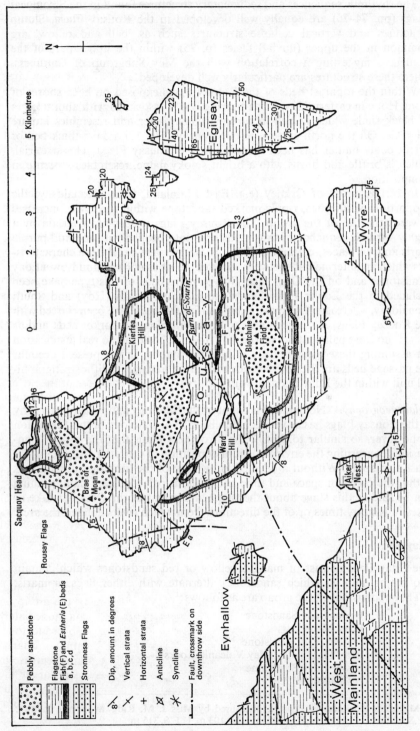

FIG. 18. *Geological map of Rousay and adjoining islands*

The rhythmic sequences and sedimentary structures found in the Stromness Flags (pp. 74–76) are equally well developed in the Rousay Flags. Slump structures and vertical collapse structures such as 'ball and pillow' are common in the upper (fluvial) facies (p. 75) within the upper part of the sequence, suggesting a correlation with the Mey Subgroup of Caithness, where these structures are particularly well developed.

Within the topmost beds of the Rousay Flags exposed on both shores of Long Hope in eastern Hoy, a fish-bearing bed is associated with about 45 cm of black shale which is rich in carbonaceous matter and resembles impure oil shale. On the north-east shore of Stronsay (Fig. 19) a 5 cm-thick bed of bright coaly matter has been recorded in the Rousay Flags. This material, which is brittle and burns with a bright smoky flame, resembles albertite in composition.

In certain areas of Orkney (e.g. East Mainland, South Ronaldsay) the proportion of siliceous, ripple-marked sandstone within the cycles increases towards the top of the group. The Rousay Flags proper are overlain by a highly variable sequence of thin sandstones interbedded with red and purple marls and, in places, calcareous flags. These 'passage beds' have characteristics which are intermediate between those of the Rousay Flags and Lower Eday Sandstones and on the Geological Survey maps beds of this type have been included in the Lower Eday Sandstone in South Walls (Hoy) and South Ronaldsay, whereas in Deerness, Eday and Sanday they have been classed with the Rousay Flags. As there appear to be no lithological marker beds and as there is no close palaeontological control, there is, however, no real justification for assuming these passage beds to be diachronous. In the present account the passage beds are for convenience treated as a separate lithostratigraphical unit within the Eday Beds (p. 81).

Palaeogeography.   No detailed account of the lithology and palaeogeography of the Rousay Flags has as yet been published. The lithology of this formation is, however, so similar to that of the Stromness Flags that the general conclusions concerning the environment of deposition of the latter (see pp. 74–77) must also apply. Without a detailed knowledge of the lithological variations within the cycles in space and time and without palaeocurrent data, nothing can be said at this stage about the shape and extent of the Orcadian lake or lakes in Rousay times or of the directions from which rivers entered the area.

Eday Beds

The Eday Beds consist of massive yellow or red sandstones which contain two major units in which sandstones alternate with either flags or marls.
The subdivisions of the group are as follows:[1]

> Upper Eday Sandstone
> Eday Marls
> Middle Eday Sandstone
> Eday Flags (with Eday Volcanic Rocks)
> Lower Eday Sandstone
> Passage Beds

---

[1]Maximum thickness in Eday, as measured by Miss J. M. Ridgway, are as follows: L.E.S. 220 m, E.F. 33 m, M.E.S. 395 m, E.M. 103 m, U.E.S. 315 m.

A. Cliffs north of Yesnaby, west coast of Orkney Mainland. Thinly bedded Stromness Flags (D1516)

Plate XI

B. Whitaloo Point, Birsay, north-west coast or Orkney Mainland. Narrow zone of folded flagstones (ruck) (C3214)

A. Sand-filled sun cracks on bedding plane of flagstone. Rousay Flags, East Walls, Hoy Orkney (D1494)

Plate XII

B. 'Horse-tooth stone', a digitate stromatolite in Lower Stromness Flags. Yesnaby west coast of Orkney Mainland (D1514)

The total thickness of the Eday Beds generally exceeds 1000 m. The lower half of the group may be equivalent to the John o' Groats Sandstone of Caithness and the entire group has been included in the upper half of the Givetian stage of the topmost Middle Devonian.

Fish remains are rare in the sandstones, but the Eday Flags have yielded a highly characteristic fish fauna which includes *Microbrachius dicki* (Plate XIV, 2) *Pentlandia macroptera* (Plate XIII, 1), *Tristichopterus alatus* (Plate XIII, 2) and *Watsonosteus fletti*. This fauna has now been recorded in a number of localities in Orkney. It closely corresponds to that obtained from the fish beds at John o' Groats. It also has some species in common with the Old Red Sandstone of south-east Shetland (p. 64).

The Eday Beds form the greater part of the island of Eday, where they crop out in the relatively unfaulted core of the Eday Syncline (Fig. 19). They also give rise to considerable, though somewhat faulted, outcrops on Sanday, Stronsay, Shapinsay, East Mainland, Burray and South Ronaldsay. The outcrops extend over a distance of 58 km from north to south, and this makes it possible to assess the lithological variations within the group in this direction.

Passage Beds

The thickness of the passage beds underlying the Lower Eday Sandstone (p. 81) varies greatly. In Eday the transition from Rousay to Lower Eday lithology is fairly rapid and obvious but in the southern isles and on Mainland there is an apparent northward and westward thinning out of the Passage Beds from South Ronaldsay (possible maximum 260 m) and Deerness towards the western part of East Mainland, the south coast of West Mainland and north Hoy.

In the northern isles the Passage Beds consist of thin, in places reddish, sandstones interbedded with beds of marl and with calcareous flags. The latter have, in several islands, yielded *Dipterus valenciennesi*. On the east coast of Deerness (East Mainland) the beds consist of thin units (up to 2 m thick) of thinly bedded purplish- or yellowish-weathering siltstone and sandstone alternating with thicker beds and lenses of massive sandstone which become more closely spaced upwards. In South Ronaldsay the Passage Beds form a thinly bedded sequence of yellow and reddish marls, flags and sandstones and on the shores of Long Hope in south-east Hoy they are made up of thin beds of hard fine-grained sandstone alternating with layers of green 'marly' flags with thin sandstone ribs.

The variation in the thickness and lithology of the Passage Beds suggests that they pass laterally by intercalation into either the Rousay flagstone or Eday sandstone facies. Such an interdigitation suggests an environment of deposition along the margin of a lake flat which was crossed by or impinged upon by the channels, flood plains and deltas of rivers. At first these distributaries were confined to small belts within the area, but later they advanced and spread out laterally to cover the entire district.

Lower Eday Sandstone

Over most of its southern outcrop the Lower Eday Sandstone consists of a bright yellow medium- to fine-grained cross-bedded sandstone, with relatively few small scattered pebbles, and with individual sets that rarely exceed 60 cm

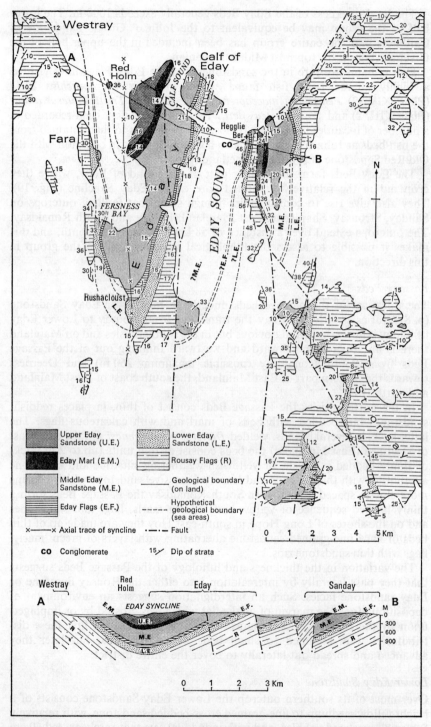

FIG. 19. *Geological map of Eday and parts of adjoining islands, showing structure of the Eday Syncline*
Section A–B across the Eday Syncline

in thickness. Complete, unfaulted, sections of the sandstone are seen in Eday, where it ranges in thickness from 100 to 220 m and consists of two distinct portions. The lower of these, which forms from half to three-quarters of the sequence, is a medium- to coarse-grained predominantly reddish purple trough-cross-bedded sandstone. It contains scattered pebbles and lenses of conglomerate which commonly form lag deposits in the bottoms of the troughs. The pebbles are up to 7 cm in diameter, usually subrounded, and consist of pink granite and pegmatite, granitic gneiss, fine-grained quartzite, beige and red chert, clear and milky vein quartz, and lesser amounts of sandstone and hypabyssal igneous rock fragments. The upper part is a yellow, medium-grained sandstone which is generally devoid of pebbles and is characterised by large-scale planar and trough-cross-bedding. In the west of Eday this sandstone is resistant to weathering and, as it splits into large rectangular blocks, it has been quarried for freestone at Fersness Bay in Eday and on Fara. Palaeocurrent data indicate that in Eday the Lower Eday Sandstone was laid down by swift currents which entered the area from the south and south-west. On Sanday the lower facies consists of pebbly coarse- to medium-grained, yellow, orange and green sandstones. The pebbles have a maximum diameter of 13 cm and are either isolated or form lags in the basal layers of cross-bedded units. The upper facies consists of medium-grained white and pale green, often mottled, sandstones with some small isolated pebbles.

As the sandstone is traced south-eastward through Stronsay into Shapinsay it becomes progressively less pebbly and predominantly yellow in colour. On Mainland good sections are seen on the south and south-east coasts of Deerness and there are extensive exposures along the north shore of Scapa Flow (Fig. 3). In Deerness the sandstone is 200 to 215 m thick, and consists of massive, predominantly yellow, medium-grained sandstone with generally small- to medium-scale cross-bedding and only rare scattered pebbles. A feature of the sandstone exposed in Deerness and many other outcrops is the high proportion of sets with some form of convolute or slumped bedding. The thickness of sandstone exposed along the north shore of Scapa Flow, on the downthrow side of the North Scapa Fault, is estimated at 550 m, which is much greater than is usual in this group. It is possible that the sandstone here includes not only the Lower Eday Sandstone but also higher members of the Eday succession. This would imply that the horizon of the Eday Flags is entirely represented by sandstone. Alternatively, the increased thickness of sediment could be due to vertical movements along the fault taking place contemporaneously with the deposition of the Lower Eday Sandstone. Tectonic activity of this kind would account for the abundance of disturbed bedding structures in the sandstone in this part of Orkney.

In South Ronaldsay the Lower Eday Sandstone consists of soft, medium- to fine-grained yellow cross-bedded sandstone with some bands of grey or yellow marly shale and siltstone and very rare pebbly lenses. The bands of fine calcareous sediment are most abundant in the southern outcrops, and on the east coast. Between Halcro Head [476 856] and Wind Wick [457 870], fish-bearing rhythmic units similar to those in the Eday Flags (p. 85) are common in the upper part of the sandstone, suggesting that this part of the sequence is perhaps better placed in the lower part of the Eday Flags. Similar rhythmic units with calcareous flags containing fish remains (usually *Dipterus*

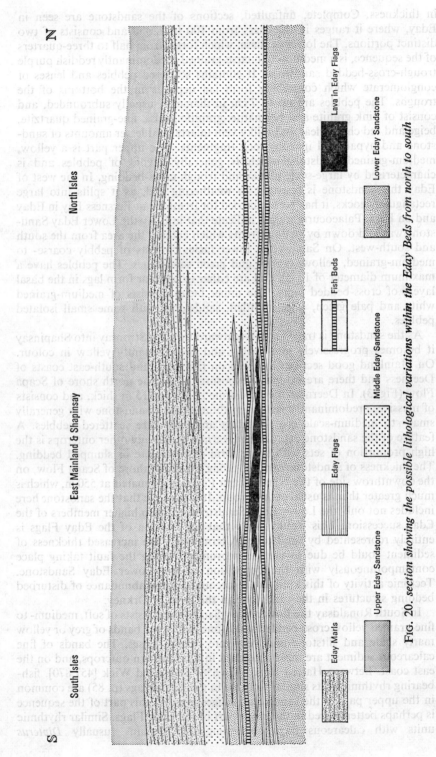

FIG. 20. *section showing the possible lithological variations within the Eday Beds from north to south*

valenciennesi, and *Pentlundia sp.*) have been recorded within the lower strata of the Lower Eday Sandstone of Flotta and these may therefore be considered as Passage Beds. In east Hoy the group consists of soft, yellow and reddish cross-bedded sandstone, and in north-west Hoy, where there are small out-crops of the sandstone at Sea Geo [261 027] on the north coast and beneath the Upper Old Red Sandstone tuffs along the west coast between Rack Wick [197 990] and Rora Head [182 992], it is characteristically soft, sulphur-yellow, fine- to medium-grained, with thin cross-bedded sets, thin bands of marly flags and rare pebbly lenses. The Lower Eday Sandstone status of the Rack Wick outcrop is, however, open to question.

Eday Flags

The Eday Flags are a group of mixed sediments of very variable thickness, which appear to pass laterally by interdigitation into the Middle and possibly also the Lower Eday Sandstone (Fig. 20). They consist of rhythmic units which range in thickness from two metres to several tens of metres. Most of these units have the following two phases:

1. A fining-upward phase of buff, yellow or, more rarely, red sandstone and sandy siltstone. This phase ranges in thickness from less than 1 m to over 25 m.
2. A coarsening-upward phase of grey, black, and locally purple 'flagstone' which generally has a finely laminated fish-bearing 'quiescent water' facies at the base succeeded by most of the other units and lithological features found in the lacustrine parts of the cycles in the Stromness and Rousay Flags.

A measured section of three Eday Flagstone cycles in Eday is shown in Fig. 21. The cycles of this formation differ in several ways from the rhythmic units of the Stromness and Rousay Flags. The sandstone phases are in many instances much thicker and the exposed sections of many cycles have one or more thick, often red, channel-fill sandstones. These sandstones in places contain scattered pebbles and they are generally coarser-grained than the sandstones in the Stromness and Rousay Flags. Within the dark calcareous finely laminated fish-bearing siltstone–shale sequences near the base of the flagstone phases, one commonly finds bands of non-laminated pale grey silty mudstones which are 10 to 30 cm thick. These non-laminated beds may be the distal deposits of intermittent turbidity currents. Sun cracks and syneresis cracks are common in most siltstones and fine sandstones, but stromato-lites have not been recorded. The thin sandstone ribs within the flagstone phase very commonly have load-cast bases, many are slumped, and some exhibit graded bedding.

The Eday Flags are thickest and best developed in and near Deerness (south-east Mainland), where they are up to 150 m thick and have a volcanic horizon near their base. Excellent exposures are seen on the west shore of Newark Bay [567 036]. In South Ronaldsay the thickness of the Eday Flags is also about 150 m, but the group contains a progressively higher proportion of sandstone as it is traced southward and westward from the north shore. In this island the flagstone phases of the cycles become more red, green and marly towards the south. Traced northwards from Deerness the formation diminishes rapidly in thickness to 100 m in Shapinsay, less than 50 m in Stronsay and south-west Eday, until in the Calf of Eday and Sanday the Lower and Middle Eday Sandstones are separated by only 10 m or so of flaggy sediments. This apparent thinning appears to be a facies change

resulting from the progressive northward thickening and coarsening of the
sandstone phases and the corresponding thinning out of the flagstone phases
in the higher cycles of the group. This process of interdigitation can be most
readily demonstrated on Eday. Thus at Rushacloust on the south-west shore

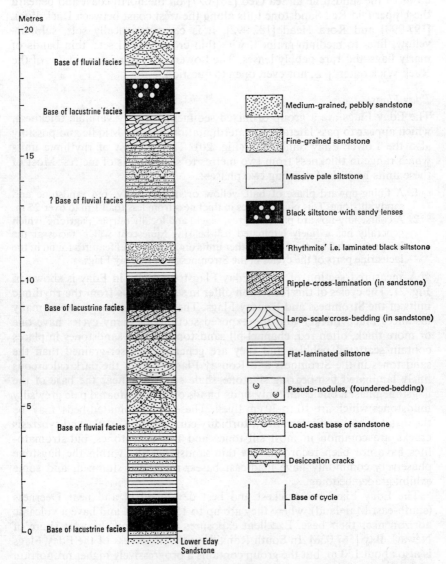

FIG. 21. *Measured section showing cyclic sequence in the Eday Flags, Rushacloust,
west coast of Eday*

of the island, the formation consists of a lower, 18 m-thick, section which
consists of three cycles with relatively thin sandstones and well-developed grey
fish-bearing flags, and an upper section, about 25 m thick, composed largely
of red, cross-bedded pebbly sandstone but containing three thin beds of

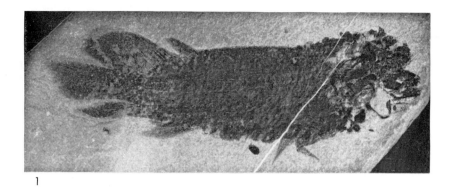

1

2

3

1. *Pentlandia macroptera* (Traquair) x1, Eday Flags
   (Royal Scottish Museum specimen)

2. *Tristichopterus alatus* Egerton x⅔, Eday Flags

3. *Osteolepis macrolepidotus* Agassiz x⅔, Upper Stromness Flags

Plate XIII

1. *Coccosteus sp.* x1½, Brindister Flags

2. *Microbrachius dicki* Traquair x2, Eday Flags
   (Royal Scottish Museum specimen).

3. *Asterolepis thule* Watson x1, Brindister Flags

4. *Millerosteus minor* (Miller) x2, Rousay Flags
   (Royal Scottish Museum specimen)

5. *Asterolepis orcadensis* Watson x1, Rousay Flags   Plate XIV

siltstone and flagstone. The number of cycles with fish beds and thin sand-stones decreases northwards to two at Fersness Bay in the centre of the island and to one on the shores of Calf Sound and on Sanday. The thin beds of flagstone in the higher cycles die out completely northwards, but on the shore of Fersness Bay an isolated fish bed crops out some 110 m above the highest remaining cycle with a flagstone phase.

Eday Volcanic Rocks

A number of isolated outcrops of fine- to medium-grained basic igneous rocks associated with thin beds of tuff have been recorded in the basal beds of the Eday Flags at Deerness, Shapinsay, and Copinsay [595 020]. The igneous rocks were interpreted as lava flows by Wilson and others (1935), but Kellock (1969) has upheld Flett's original concept (Flett 1898b) that the rocks exposed at Muckle Castle (Deerness) and the Black Holm of Copinsay are intrusions. The former is probably a volcanic plug and the latter a sill. The basalts of Haco's Ness (Shapinsay), Point of Ayre (Deerness), and the inland exposures in Deerness are lava flows with scoriaceous tops, pipe-amygdales and sand-filled fissures. At least two flows are present and though one is very thin the other is at least 7 m thick. The intrusive rocks and the lavas are alkaline olivine-dolerites and basalts which contain interstitial analcime, natrolite and alkaline feldspar. They are petrographically similar to the Carboniferous teschenites and basanites of the Midland Valley of Scotland. The field and petrographic evidence, however, leaves no doubt that the lavas and intrusions are coeval and that both are of Middle Old Red Sandstone age.

Middle Eday Sandstone

The Middle Eday Sandstone ranges in thickness from possibly 400 m in Eday and Sanday to only about 90 m in East Mainland and South Ronaldsay. This southward decrease in thickness appears to be due to the lateral passage of the sandstone facies into both Eday Flag and Eday Marl type lithology (Fig. 20). Within the beds mapped as Middle Eday Sandstone there is a marked reduction in overall grain-size and an increase in the proportion of marly sediment from both the north and the south towards Deerness. In Eday and south-west Sanday the sandstone consists essentially of reddish purple, principally trough-cross-bedded, medium- to coarse-grained gritty sandstone with scattered pebbles and conglomerate lenses. Conglomerates are particularly well developed at Hegglie Ber on the south-west coast of Sanday. The pebbles consist predominantly of coarse granite, quartzite and vein quartz but throughout the entire thickness of the sandstone and in the over-lying groups there are also an appreciable number of rounded clasts of porphyritic and spherulitic rhyolite and scoriaceous basic lava. A few silty and marly beds are present in the lower part of the sequence.

In Stronsay the formation is composed of red and yellow sandstones, with only a few scattered pebbles and with marly beds, which are particularly prevalent towards the base. In Deerness the only undoubted outcrop of the Middle Eday Sandstone is that exposed at Newark Bay [569 040] where the strata immediately above the Eday Flags consist of a series of fining-upward cycles, each made up of a sandstone unit 50 cm to 2 m thick overlain by deep purple poorly laminated sandy siltstone and siltstone with thin ribs of com-monly convoluted sandstone. Apart from their duller purple colour these

G

beds are similar to the typical sediments of the Eday Marls (p. 88). The sequence exposed at Newark Bay appears to pass upwards into more massive sandstones. West and south of Deerness, at Scapa Bay and Burray, the higher sediments of the formation are exposed, these being red, white, pink and yellow sandstones with thin marly bands. Still further south, in South Ronaldsay, the sediments ascribed to this formation consist largely of red and yellow cross-bedded sandstones with only rare pebbles and with thin bands of grey, red and yellow 'marl'.

Eday Marls

This formation consists principally of thick beds of bright red and pale green calcareous sandy siltstone and pale green siltstone, alternating with thinner beds of hard, cross-bedded red and yellow sandstone. The sandstone bands and lenses become thicker and more closely spaced in the upper part of the sequence. The thickness of the group is over 200 m in South Ronaldsay, at least 150 m in East Mainland and then decreases northwards to about 100 m in Eday and even less in Sanday. Again, much of this thinning may be due to the interfingering of the marl facies with the adjoining sandstone facies (Fig. 20).

The diagnostic feature of the formation is the bright red colour of the fine-grained sediments. The rhythmic units with their high proportion of marly siltstones are fining-upward cycles of the type generally thought to be formed by meandering streams with shifting channels and extensive overbank deposits. In Eday the channel sandstones in the lower part of the sequence are less than 2 m thick and are separated from each other by up to 15 m of red sandy micaceous siltstone which is poorly sorted and has some ripple marks, sun cracks and traces of bioturbation. There are also bands and zones with calcareous concretions. As the sequence is followed upward the ratio of sandstone to marl increases and there is a gradual passage into the overlying Upper Eday Sandstone. The Eday Marls also crop out in East Mainland where they form the Head of Holland Peninsula [490 120] just east of Kirkwall, as well as in Burray, Hunda [432 969] and the north-west of South Ronaldsay. In these areas the beds form cyclic units which are similar to those in the lower part of the Eday Marls sequence of Eday.

Upper Eday Sandstone

The Upper Eday Sandstone crops out in two areas: a northern area which comprises the outcrops in Eday and south-west Sanday, and a southern area around the south shore of Scapa Flow, which includes the outcrops on Burray, north-western South Ronaldsay and Flotta. In the southern area the group consists of beds of soft, cross-bedded, red, pink and yellow sandstone alternating with bands of red and green 'marl'. The latter are most common and thickest at the base of the formation. In Eday, where the group reaches a thickness of more than 300 m, it falls into two lithological divisions. The lower division consists of beds of red and yellow sandstones with scattered pebbles, which pass upwards into more massive trough-cross-bedded pebbly sandstones alternating with beds of marl. The pebbles and sedimentary structures in the latter are closely comparable to those found in the Middle Eday Sandstones. The upper division shows a reversion to finer-grained sediments and consists of bright red and purple sandstones with thin bands of sun-cracked sandy marl.

Conditions of Deposition

No modern account of the lithology and palaeogeography of the Eday Beds has as yet been published, and the following suggestions as to the environment of deposition and direction of palaeocurrents are based on observations made by the author and Dr N. G. T. Fannin during a brief spell of field work in 1972.

The sandstones and red 'marls' of the Eday Beds were deposited in the channels, alluvial fans and flood plains of fairly large rivers which entered the area from the south-west and either filled up the Orcadian lake or at least encroached upon its margin. The presence within the Eday Flags of grey thinly bedded lacustrine flagstones, which alternate with channel sandstones and lake delta deposits, indicates that there was a time when the waters of a lake again covered the greater part of the Orkney area. This lake was fed by swifter and more active rivers than those which entered the lake at the time the Stromness and Rousay flags were laid down. They regularly built out deltas into the lake and these pushed its margin north-eastwards beyond the limits of the present area. The lithology of the higher beds of the Eday group indicates that a fluvial regime was soon re-established, first in the north, but then over the whole area. The Eday Sandstones were probably laid down by swift, braided and straight rivers which formed alluvial fans, but the Eday Marls were formed in the channels and alluvial plains of slower meandering streams. The appearance of swift-flowing rivers is probably the result of differential vertical movements along faults which were responsible for the tectonic uplift of the source area or areas.

Upper Old Red Sandstone

The only Orkney rocks classed as Upper Old Red Sandstone are the thick sandstones and underlying volcanic rocks which occupy the greater part of western Hoy. The latter rest on an irregular eroded surface floored by faulted and gently folded Upper Stromness Flags and Lower Eday Sandstone. The sandstone of Hoy has been equated with the Dunnet Head Sandstone of Caithness, as the two have identical lithological characters. Neither had, at the time of writing, yielded any diagnostic fossils[1] and their suggested Upper Devonian age was based solely on the fact that a phase of tectonic activity, followed by a lengthy period of erosion, intervened between the deposition of the Middle Old Red Sandstone sediments and the eruption of the Hoy volcanics. The results of radiometric age dating of the Hoy lavas are somewhat ambiguous, but it is now thought that the most likely age of these rocks is around 353 million years, which is Upper Devonian.

---

[1] Scales of *Holoptychius*, a characteristic Upper Old Red Sandstone fish, have since been found in the Dunnet Head Sandstone, Caithness, by Mr. A. McAlpine (personal communication).

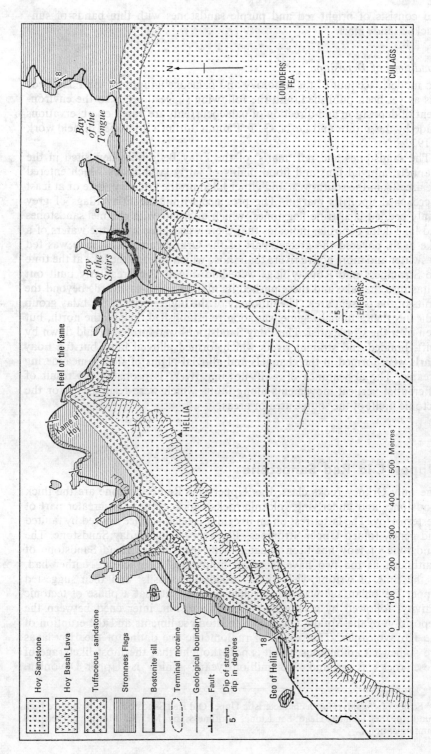

FIG. 22. *Geological map of the north-west corner of Hoy*

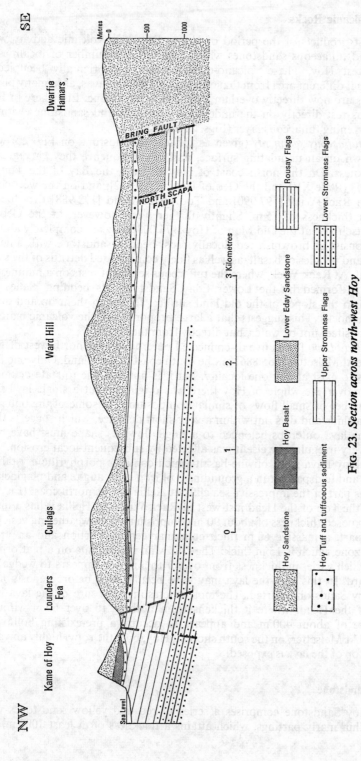

FIG. 23. Section across north-west Hoy

The position of the Kame of Hoy and Cuilags is shown on Fig. 22, the present section continues further to the S.E.

NW

SE

Kame of Hoy

Lounders Fea

Cuilags

Ward Hill

Dwarfie Hamars

BRING FAULT

NORTH SCAPA FAULT

Sea Level

Metres
0
500
1000

0 1 2 3 Kilometres

Hoy Sandstone

Hoy Basalt

Hoy Tuff and tuffaceous sediment

Lower Eday Sandstone

Upper Stromness Flags

Rousay Flags

Lower Stromness Flags

Hoy Volcanic Rocks

The first products of the period of Upper Devonian volcanic activity were tuffs and tuffaceous sandstones which crop out in a number of localities in north-west Hoy. These volcano-detrital sediments originally occupied a somewhat different area from that covered by the later lavas, as in many places the tuffs are now directly overlain by the Hoy Sandstone. Elsewhere in Hoy the lavas rest directly on the underlying flagstones and sandstones without an intervening tuffaceous layer (Figs. 22 and 23).

The *tuffaceous sediments* (given as tuffaceous sandstone on Fig. 22) were laid down on an undulating surface, and in consequence they vary greatly in thickness. On the north coast of Hoy between the Bay of the Tongue [207 047] (Plate XV) and the Geo of Hellia [190 042], and on the west coast between Rack Wick [197 990] and Too of the Head [192 988] they locally reach a thickness of 15 m. Slightly farther west, however, at the Geo of Hellia itself and at the Old Man of Hoy [176 008] they are completely absent. They consist of brownish red, locally cross-bedded sandstone with angular blocks and pebbles of basalt, as well as finely comminuted detritus of the same material. At Rack Wick, where the tuffaceous sediment rests on a hummocky basement formed by the Lower Eday Sandstone, its bedding planes are parallel to the slopes of the old land surface, following them up and down the hummocks. This suggests that a large proportion of the volcanic material was deposited, not by water, but directly from the air.

The *Hoy Lava* forms five disconnected outcrops in the north-west of Hoy and one at Melsetter [265 886] on the south coast of the island. Only one flow appears to exist at any one locality, but it is not possible to state definitely whether over the whole of Hoy there was originally but a single lava flow or a number of small flows of similar composition. In some of the outcrops the lava thins and dies out within a very short distance, which suggests that, if all the lava outcrops belonged to a single flow, its shape must have been either very irregular, or else it was affected by subsequent local erosion.

The Hoy Lava is an olivine-basalt which contains porphyritic crystals of olivine and feldspar, set in a groundmass of iron ores, augite and plagioclase. It forms part of the impressive sea cliffs at Hellia on the north coast (Fig. 22) and at the Too of the Head just west of Rack Wick. At Hellia it has a maximum exposed thickness of about 90 m comprising a grey-weathering vesicular lower part, a massive 60 m thick columnar central portion and an upper slaggy zone which is 15 m thick. The Old Man of Hoy sits on a platform of lava, which varies in thickness from over 7 to 3 m and appears to wedge out westward. In this area the lava may have been eroded before deposition of the Hoy Sandstone started. The most interesting exposure of the lava is at Too of the Head, where it thickens from nothing to over 60 m within a distance of about 400 m and appears to occupy a pre-existing hollow or valley. At Melsetter, on the south shore of Hoy, only the ropy, highly amygdaloidal top of the lava is exposed.

Hoy Sandstone

The Hoy Sandstone comprises a series of red and yellow sandstones with some thin marly partings, which attains a thickness of at least 1000 m. No

Kame of Hoy and Bay of the Stairs, north-west Hoy, Orkney
From top to base 1. Hoy basalts (black, jointed) 2. Hoy tuffaceous sandstone (red-brown) 3. Upper
Stromness Flags (thinly bedded, forming greater part of cliff), 4. Bostonite sill (base of cliff) (D1492)

Plate XV

remains of fish or plants have as yet been found in these beds, though Stephens and Edwards (*in* Wilson and others 1935, pp. 140–41, fig. 20) have recorded the presence of rather doubtful 'tracks left by some creature as it crawled about over the mud' in blocks of sandstone in the Burn of Redglen [219 014] on the west side of Ward Hill.

The Hoy sandstones are medium-grained, red or yellow in colour, and generally trough-cross-bedded, with individual sets ranging to little more than 1·2 m in height. Slumped cross-bedding and convolute-bedding are common at certain horizons. Many sets contain chips of red and purple siltstone, and intraformational conglomerates at the bases of troughs are common. There are also rare small lenses of extraformational conglomerate as well as scattered pebbles of quartz, schist and gneiss. In a limited sector of the south shore of Hoy, where the author has recorded palaeocurrent data, the sandstones appear to have been deposited by currents moving to the east or east-north-east, but these findings do not necessarily apply to the entire group. It is likely that most of these beds were laid down in a fluvial environment, probably by braided rivers.

There appears to be some variation in the lithology and the resistance to weathering of the sandstone in different parts of the sequence. Thus in the hills of north Hoy and on the cliffs of the north shore the lower beds of the Hoy Sandstone are soft and friable and in places weather readily into loose sand. The overlying beds are much harder and form the prominent crags along these hillsides. The presence of the softer layers overlain by harder sediments has thus given rise to the steep-sided outlines of the hills in this part of Hoy. The near-horizontally bedded sandstone with its vertical joints has been eroded by the sea into the spectacular vertical cliffs and seastacks of western Hoy. The Old Man of Hoy (cover photo) is a vertical rectangular pillar of sandstone which towers to a height of 137 m above sea level. Almost equally spectacular is the rectangular pinnacle of rock just south of St John's Head [187 035], which has been almost separated from |the main cliff by movement along the vertical joint planes. Further to the south-east, along the shore south of Little Rack Wick [233 932], the vertical sandstone cliffs are undercut by closely-spaced rectangular sea caves.

REFERENCES

Crampton and Carruthers 1914; Donovan and Foster 1972; Fannin 1969, 1970; Flett 1897, 1898b; Kellock 1969; McQuillin 1968; Miles and Westoll 1963; Richardson 1965; Ridgway 1974; Steavenson 1928a, 1928b; Lankester and Traquair 1868–1914; Westoll 1937, 1951; Wilson and others 1935.

9. MINOR INTRUSIONS

Shetland

Western Shetland

Swarms of sub-parallel roughly north–south trending acid, intermediate and basic dykes cut the plutonic complexes and adjoining metamorphic and sedimentary rocks of North Roe, Northmaven, Muckle Roe and the northern part of the Walls Peninsula (Fig. 24). The acid dykes comprise feldspar-phyric porphyrites, quartz-feldspar-porphyries and a great variety of felsites which are generally either flow-banded or spherulitic. In North Roe there is a suite of felsite dykes which contain the sodic minerals riebeckite and aegirine. These rocks have a striking blue or bluish green colour and are, in many instances, spherulitic. The thickness of individual acid dykes ranges from a few centimetres to about 18 m. A few have lenticular outcrops and some of these are locally up to 55 m wide. The basic dykes are principally basalts and dolerites, but many of them contain hornblende as the dominant dark mineral instead of pyroxene. A great deal of the hornblende is secondary and may have formed during a late hydrothermal period of uralitisation. Some of the basic dykes contain appreciable quantities of quartz and these have been classed as quartz-dolerites. There are also a small number of thick dykes of coarse pyroxene-porphyrite. Some of these have a highly irregular outcrop. Dykes of intermediate composition are relatively rare; they include kerato-phyres, spessartites and microdiorites.

Both the basic and acid dykes generally form parallel swarms with trends ranging from north-north-west to north-north-east. A few of the dykes have curving courses within that sector. Exceptions to this general north–south trend are found on Vementry Island, where acid dykes tend to radiate from the Vementry Granite and in the northern part of the Walls Peninsula where some acid and basic intrusive bodies trend parallel to the strike of the Old Red Sandstone sediments. In the ground extending from North Roe to Muckle Roe the acid dykes are restricted to a north–south belt which coincides in width with that of the outcrops of the granites and granophyres. The distribution of the riebeckite-bearing felsites is even more restricted to a median north–south trending zone in and adjoining the Ronas Hill Granite. The basic dykes, on the other hand, are fairly evenly distributed over a rather wider area. On Muckle Roe acid and basic dykes form alternate swarms. Composite acid–basic dykes are relatively rare, but a number of acid dykes have keratophyric margins.

Most acid and basic dykes are younger than the plutonic masses which they cut, but not all dykes are younger than the latest granitic members of the complex. There is evidence for the existence of some basic dykes which are earlier than the adjoining granite. In North Roe the evidence of chilling of adjoining dykes and the intersection of dykes has shown that in most cases

94

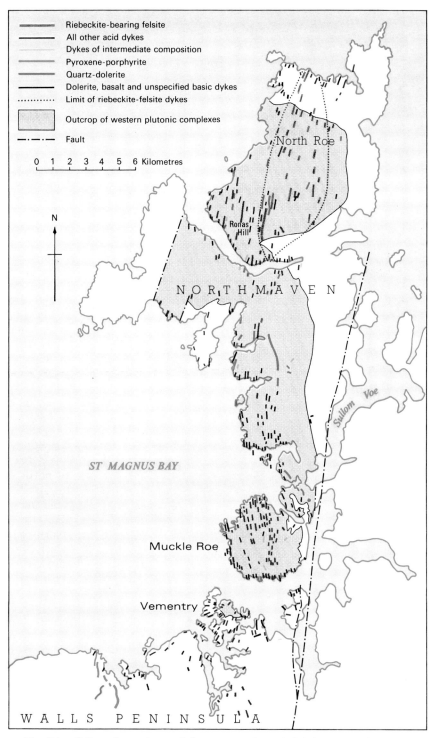

Key:

- ▨▨▨▨▨ Riebeckite-bearing felsite
- ——— All other acid dykes
- ——— Dykes of intermediate composition
- ——— Pyroxene-porphyrite
- ——— Quartz-dolerite
- ——— Dolerite, basalt and unspecified basic dykes
- ·········· Limit of riebeckite-felsite dykes
- ▨ Outcrop of western plutonic complexes
- —·—·— Fault

0 1 2 3 4 5 6 Kilometres

N

North Roe

Ronas Hill

N O R T H M A V E N

Sullom Voe

ST MAGNUS BAY

Muckle Roe

Vementry

W A L L S   P E N I N S U L A

Fig. 24.   Minor intrusions of the west and north Mainland of Shetland.

the basic dykes, spessartites and microdiorites are older than the quartz-feldspar-porphyries and these are older than the felsites. The blue riebeckite-felsites are the youngest of all. Further south the evidence for the age relationships of the dykes is less clear, and in the Walls Peninsula the small number of dyke intersections show basic dykes cutting acid ones. It is believed that in Northmaven and Muckle Roe the emplacement of acid and basic dyke-swarms proceeded in a series of alternating pulses. The close compositional and areal relationships between the late Caledonian plutonic complexes and the dyke swarms suggests that the dykes of Western Shetland are also of late-Caledonian and possibly Upper Devonian age.

In contrast to the northern plutonic complexes, the Sandsting Complex (Fig. 9) has few minor intrusions either cutting it or associated with it. There are a number of small dykes of melamicrodiorite, some of which have the aspect of uralitised dolerites. Felsite dykes are rare within the complex and in the sandstones adjoining the granite, but on the shores of Bixter Voe [33 51] and Gruting Voe [27 48] there are dyke-swarms of granophyre and graphic microgranite, which appear to have been derived directly from the granitic magma.

Eastern Shetland

Two suites of minor intrusions are present in the metamorphic rocks of east Mainland but, with the exception of one trachytic dyke, there are no igneous intrusions in the eastern Old Red Sandstone. The earlier suite consists of dykes which are metamorphosed and are mainly microdiorites, lamprophyres, and more rarely felsic porphyrites. These dykes possess a schistosity which is parallel to their margins, but does not continue into the country rock. It is thought that the dykes were deformed immediately after their intrusion and that they acted as movement zones in which differential slip gave rise to the schistosity and probably accelerated the process of crystallisation.

The second suite of minor intrusions comprises unmetamorphosed dykes and sills. They have been loosely grouped as lamprophyres, mainly spessartites and while most are not more than a few metres thick, some can be followed along their length for hundreds of metres. Though these dykes are generally unaffected by metamorphism they are sheared and even schistose in areas where they are involved in shear zones, such as those separating the individual tectonic blocks of Unst and Fetlar (p. 32). 'Lamprophyre' dykes are particularly common in the Graven Complex of northern Delting (Fig. 8) where they appear to have been intruded before the emplacement of the 'Inclusion Granite' (p. 42).

Sheets of quartz-porphyry cut the metamorphic country rock in the Channerwick district of south Mainland (Fig. 8), where outcrops occur both north and south of the Channerwick Granite (p. 45) and the sheets are clearly associated with the plutonic mass.

Foula

The metamorphic rocks exposed along the east coast of Foula are cut by a network of sills and dykes of pink porphyritic microgranite. The sills are thickest and most abundant near the northern end of the outcrop where one thick sill forms the 650 m-long stretch of foreshore at Ruscar Head (Fig. 12). The groundmass of the microgranite is slightly sheared and granulitised. None of the granite intrusions cut the Old Red Sandstone of Foula, suggesting that the microgranite was emplaced prior to its deposition.

Fair Isle

The west and south coasts of Fair Isle are cut by a number of dykes of partly uralitised dolerite and basalt. Both porphyritic and ophitic dolerites are present. At three localities the basic dykes are closely associated with dykes of microgranite and felsite. Most dykes are emplaced along west-north-west trending crush belts (Fig. 15). Sodic scapolite, together with calcite, and locally analcime, apatite and sphene form veins within and close to the basic dykes and the crush belts. Copper ores and pyrite are associated with some of these veins.

The Fair Isle dykes are similar to the late-Caledonian dykes of western Mainland, and the scapolite mineralisation appears to be connected with a plutonic complex which may lie a short distance offshore and which may be related to the plutonic complexes of west Shetland.

Orkney

Over two hundred dykes and a small number of sills have been recorded in the Orkney Islands. Most of the dykes are of dark lamprophyric rocks which are classed as camptonites and monchiquites, but there are also a small number of highly feldspathic dykes and sills which have been termed bostonites. Dykes of these three types do not occur in Shetland and only a few have been recorded in northern Caithness. Even in the Orkney Islands they have a restricted distribution and are mainly confined to the south and west of the archipelago (Fig. 25). The camptonites are most common in the West Mainland and in Rousay; the monchiquites in the south islands. The trend of the camptonites is predominantly east-north-easterly, that of the monchiquites is north or north-north-east. It seems most likely that the dykes came from a focus that lies somewhere to the south or south-west of the island group.

The field relationships of the dykes indicate that they are younger than the faults which cut the Old Red Sandstone. Recent radiometric potassium-argon dates by N. J. Snelling (personal communication) have suggested that their most likely age is around 283 ± 9 million years, which is late-Carboniferous. This accords with the ages of similar dyke suites in the West Highlands and the Midland Valley of Scotland. There is no evidence that any of the dykes ever gave rise to surface lava flows. On the other hand, some of the small volcanic vents or cryptovents in the south-east of the island group (p. 103) are either intimately connected with monchiquite dykes or contain clasts or lapilli of basic igneous, possibly monchiquitic material. This suggests that the formation of these vents was connected with the intrusion of monchiquite magma.

Two intrusions of alkaline olivine-dolerite in Deerness (East Mainland) and on the Black Holm of Copinsay, which are intimately connected with the volcanic rocks in the Eday Flags, are described on p. 87.

FIG. 25. *Minor intrusions and volcanic vents of Orkney*

Camptonites. The camptonites are the most numerous of all the Orkney dykes and they consist of phenocrysts of olivine, augite and hornblende set in a groundmass composed of augite, hornblende, feldspar and iron ores. Porphyritic olivine is most common in the finer-grained dykes, but it is hardly ever found in an unaltered state. The augite phenocrysts generally form near-euhedral crystals which have a deep green core and a pale mauve or lilac outer zone when seen in thin section. The hornblende, when present, is generally of a deep brown colour in thin section, and the larger hornblende crystals may have rounded edges.

In the groundmass near-euhedral augites and dark brown hornblendes are abundant. The feldspars, which are generally zoned, with labradorite in the centre and oligoclase or alkali-feldspar on the margins, fill the spaces between these earlier minerals. A characteristic feature of the groundmass of the camptonites is the presence of numerous light-coloured rounded ocelli which may represent steam cavities partly filled by alkali-feldspar during a later crystallisation. Many of these ocelli have a central residual cavity which is filled with carbonates or analcime. Most of the camptonite dykes are in a highly decomposed condition and the freshest material can be obtained from pebbles in boulder clay or shore gravel.

The most remarkable camptonite dykes of Orkney are those exposed at Hoxa in South Ronaldsay. These are crowded with phenocrysts of augite and hornblende which are up to 12·5 mm in diameter.

Monchiquites. The monchiquites are rocks which consist mainly of olivine and augite. The former forms microphenocrysts which are up to half a milli-metre in size and which are in many cases altered to serpentine, chlorite, and carbonates. The augite makes up about two-thirds of the rock and forms small near-euhedral crystals. Brown hornblende is a minor constituent of some monchiquites, and small plates of biotite are present in most dykes. In the freshest rocks the groundmass is a brown glass, perhaps analcimic in composi-tion, with curving microlites and small grains of iron ore. The original groundmass is, however, rarely preserved and in most dykes it is replaced by an aggregate of carbonates with serpentine, chlorite, zeolites and a good deal of analcime. Ocelli are not as common as in the camptonites. When present they contain some crystals of feldspar together with hornblende, biotite and nepheline.

Most monchiquites are less coarsely porphyritic than the camptonites, but there are some monchiquite dykes with conspicuous phenocrysts. A dyke on Corn Holm, Copinsay, for instance, has crystals of augite, hornblende and olivine which are more than 2·5 cm in diameter.

In about twenty of the monchiquite dykes fresh nepheline has been found to occur as a late crystallisation product in the groundmass or in the outer parts of certain ocelli. These nepheline-monchiquite dykes have no separate area of distribution but occur among and along with the other monchiquites. Though some biotite is present in most monchiquite dykes, there are a few dykes with ophitic flakes of golden-yellow biotite which are up to 1 mm across. These biotite flakes are commonly associated with olivine. Other minerals present in the biotite-monchiquites are granular carbonates, augite, perov-skite, apatite, iron ores and spinel. The biotite-monchiquites have a close affinity with the alnöites.

Bostonites. Bostonite intrusions are much rarer than those of the two rock types described above. The two finest examples are a broad dyke at the Haven, Swona and a sill which cuts the Stromness Flags just beneath the Hoy Volcanic Rocks at the Bay of the Stairs in north-west Hoy (Fig. 22, Plate XV). The Swona bostonite consists of almost pure feldspar, principally orthoclase, and is one of the most potash-rich igneous rocks in Britain. The sill on Hoy is similar in composition but contains, in addition to feldspar, some chlorite and carbonate.

Some dykes in Stenness and at the Loch of Skaill, West Mainland, are transitional in composition between bostonites and camptonites.

Olivine-basalts. A small number of dykes of olivine-basalt have been recorded in Orkney. They crop out at Firth, near Finstown, and at Loch of Harray and differ from the camptonites in that they contain porphyritic plagioclase feldspars and no brown hornblende. They do not bear a close resemblance to the Tertiary basalt dykes of Western Scotland and are probably of the same age as the Orkney camptonites and monchiquites.

REFERENCES

Flett 1900; Flinn 1954, 1967a; Kellock 1969; May 1970; Miller and Flinn 1966; Mykura 1972c; Mykura and Phemister 1976; Peach and Horne 1884; Phemister and others 1950; Phillips 1926; Pringle 1970; Read 1934c; Wilson and others 1935.

10. VOLCANIC BRECCIAS AND VENTS AND ASSOCIATED MINERALISATION

Shetland

Bressay and Noss

Two north–south trending belts of steeply inclined and locally inverted strata traverse the gently inclined Middle and ?Upper Old Red Sandstone strata of Bressay and Noss (Fig. 26). The western 'steep belt' is up to 800 m wide and contains extensive irregular outcrops of breccia. The breccia consists of blocks of sediment of the type forming the surrounding non-brecciated areas, set in a matrix of finely comminuted sedimentary material of sand and silt grade. Individual blocks range in diameter from a few centimetres to tens of metres and some enclaves have a diameter of up to 200 m. Those less than 2 m in size are subangular to subrounded and randomly rotated. The larger blocks and enclaves usually retain their original orientation, though some are tilted, so that the bedding in them is now almost completely inverted. Some of the blocks are pierced by one or more generations of veinlets of intrusive tuffisite composed of fine sedimentary detritus, and a few contain crypto-vents of intrusion-breccia.

The eastern steep belt, which traverses Noss, has no major breccia masses actually within it. Immediately west of it, however, there is a faulted syncline and this contains a large mass of sediment-breccia which forms the western peninsula of Noss and a dyke of altered ?aegirine-trachyte exposed along the north-west coast of the island. On the east side of Bressay, near the axial trace of the syncline, there are several breccia vents and one sill of intrusive tuffisite. One elongate fissure-vent exposed near Muckle Hell (Fig. 27), on the east coast of Bressay, has dykes of flow-banded carbonate rock (containing up to 95 per cent carbonate, mainly ankerite and calcite, as well as small clasts of sediment) intruded along its margins. These dykes also cut the vent breccia and the adjoining country rock. Their origin is not certain; they may have formed as intrusive tuffisite which has been largely replaced by carbonate, or they may be the product of a viscous carbonate magma. The latter origin is more likely. The breccias and associated sediments of both steep belts contain veins of crystalline carbonate and quartz with small quantities of iron and copper sulphides.

The two steep belts form the near-vertical limbs of monoclines in the sedimentary cover of this area, and they probably reflect the presence of north–south trending fracture belts in the metamorphic basement. Volcanic gases, carbonate-rich aqueous solutions and some magma appear to have passed upwards from the fracture belts into the sediment. In the western steep belt the gases penetrated along irregular cracks in the already shattered sedimentary rock and, by a process of gas-fluxion rather than explosion, they

100

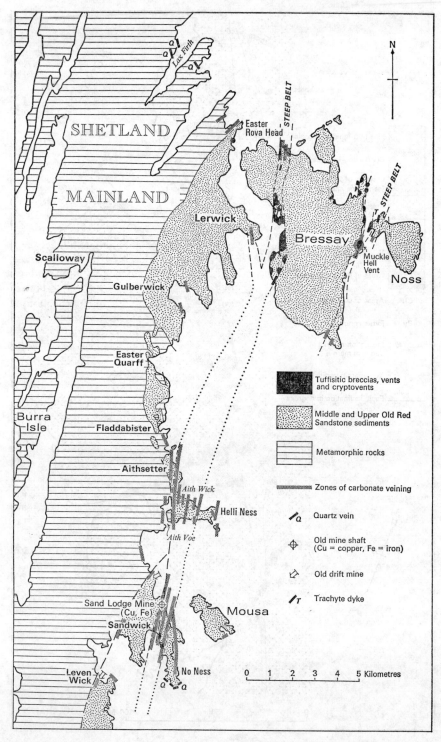

FIG. 26. *The 'steep belts', breccias, and zones of carbonate mineralisation in south-east Shetland*

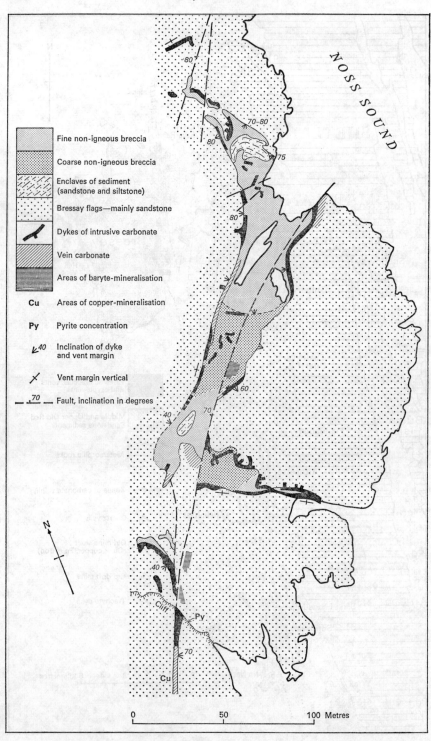

Fig. 27. *Muckle Hell Vent, Bressay (for location see Fig. 26)*

produced the large irregular areas of breccia. There appears to have been no significant upward or downward movement of solid material within the breccia. The eastern, westward sloping, steep belt did not appear to act as a funnel for escaping gases in the same way. Here the gases seem to have escaped directly upwards from the fissure into relatively unbroken rock and so gave rise to well-defined vents, dykes and sills. Most of the vents probably did not reach the surface.

South-east Shetland

The western steep belt of Bressay can be projected southwards to connect with a 500 to 1200 m-wide belt of intensive faulting and carbonate mineralisation which crosses the Helli Ness and Sandwick peninsulas of south-east Mainland (Fig. 26). This belt contains several thick dyke-like masses of carbonate, extensive areas of carbonate net-veining, some narrow and irregular zones of tuffisitic breccia and one fault-bounded horst of the metamorphic basement. The carbonate veins within and close to this belt are in some cases the gangue of ore minerals which include pyrite, hematite, chalcopyrite, malachite, gersdorffite and niccolite. Copper and iron ores have, in the past, been commercially exploited at Sand Lodge and Setter mines (p. 117). At Sand Lodge the strata in the shatter belt containing the veins have a monoclinal structure similar to that in west Bressay. Carbonate veins with traces of ore minerals are found all along the east coast of Mainland between Nesting and Levenwick, and the presence of similar veins in Fair Isle suggests that mineralisation of this type continues well to the south of Shetland Mainland.

Chemical analysis has shown that the carbonate dyke and vein rocks do not contain the trace elements, such as rare earths and niobium, which are generally present in carbonatites. Instead they have high trace concentrations of arsenic, nickel, chromium and copper. It is thought most likely that the gases and carbonate-rich solutions responsible for the breccia and vein formation were generated by the contamination of a magma by limestone and magnesite within the metamorphic basement.

Orkney

Several small volcanic vents have been mapped in the area which extends from south-eastern Hoy and South Ronaldsay north-eastward to Deerness in south-east Mainland (Fig. 25). There is also a cluster of vents and volcanic plugs in northern Hoy. In addition, a number of small crypto-vents have recently been recorded at Harra Ebb on the west coast of Mainland.

The breccia-vents in south-east Orkney consist of two types:

1. Those composed largely of clasts of basic igneous material (usually monchiquite), set in a matrix of basic tuff;
2. those consisting principally of angular fragments and blocks of flagstone and sandstone set in a matrix of ground-up sediment which in some vents contains some small rounded lapilli of basic igneous material.

The former are always small (less than 20 m in diameter) and fairly regular in outline. They usually contain some clasts of corroded and partially absorbed sediment. The latter generally have an irregular shape and vary greatly in size. The most interesting of these vents is the irregular sediment-breccia exposed on the foreshore at Croo Stone at the north-east corner of South Ronaldsay.

H

This measures 160 m from north to south and about 90 m from east to west and has dykes of monchiquite (locally up to 4 m wide) intruded both along its margin and within the vent. Another example is the small but spectacular vent seen on the cliffs west of Green Vale on the west coast of the same island. This has a carbonate-rich matrix of comminuted sediment but has no monchiquite dykes directly connected with it. Most of the sediment-filled vents of this area, however, have some monchiquite associated with them and it must be assumed that their formation was in some way connected with the uprising of monchiquitic magma. It is probable, that, like the sediment breccias of east Shetland, the vents were formed by a process of gas fluxion and they may never have extended to the surface. Flett has suggested that the gas (mainly steam) may have been generated when the ascending magma met with local concentrations of ground water.

At Harra Ebb near Yesnaby (West Mainland) (Fig. 25) there are six small oval-shaped cryptovents, none of which exceed 11 m in diameter, and one fissure vent which can be intermittently traced for a distance of 50 m. All these contain only angular to subrounded sandstone clasts of varying size, set in a matrix of comminuted carbonate-rich sandstone. Though a camptonite dyke crosses the area within which these vents occur, there is no direct evidence to connect them with the camptonite magma. Irregular masses of sediment breccia and narrow belts of fissure breccia, usually associated with small faults and monoclinal flexures, have been found farther south on the West Mainland coast and elsewhere in Orkney. In some areas these fissure breccias contain galena or chalcopyrite.

Five volcanic orifices have been recorded near Breibister in northern Hoy. These comprise two plugs formed entirely of basalt, two vents which contain both tuff and basaltic material and one vent which consists entirely of breccia. All these vents and plugs cut the Upper Stromness Flags which are here overlain by the Hoy Volcanic Rocks. As no vents have been recorded in the Hoy Sandstone that crops out a short distance to the south it has been assumed that these orifices were the feeders for the overlying Hoy lavas and tuffs.

REFERENCES

Dron 1908; Macgregor and others 1920 (pp. 217–20); Mykura 1972a, c; Peach and Horne 1884; Wilson 1921 (pp. 147–151); Wilson and others 1935.

II. PLEISTOCENE AND RECENT[1]

Shetland—Main Island Group

The first comprehensive account of the glaciation of the Shetland Islands was produced by Peach and Horne (1879b) who concluded that during the 'primary glaciation' Shetland was overridden by an ice sheet which originated in Scandinavia and approached the island group from the north-east. As it crossed the axis of Shetland its course was deflected to the north-west, probably by the pressure of ice moving northwards from the Scottish mainland. They showed that, after the pressure of the Scandinavian ice had eased a local ice cap was established over the island group and that this flowed off the land in all directions. They recognised a final phase of glaciation, when small local glaciers occupied the valleys and left numerous moraine heaps sprinkled over the area. Peach and Horne also drew attention to the absence of 'kames' (i.e. the deposits left by glacial meltwaters) and the complete absence of raised beaches.

Support for the presence of Scandinavian ice in Shetland was provided by the discovery in 1900 of a large glacial erratic weighing over 2 tons at Dalsetter in south-east Mainland (Fig. 28). This consists of tönsbergite, a rock known to occur in situ only in southern Norway (Finlay 1926a).

Since Peach and Horne's pioneering studies there has been much additional information about the glaciation of these islands, and their conclusions have been considerably modified. It is recognised that Shetland, like the rest of Scotland, must have been covered by ice during all four glacial maxima of the Pleistocene Period, but that most of the superficial deposits and ice-formed features are attributable to the last (Devensian or Weichselian) glacial episode.

The ice sheets have greatly modified the Shetland landscape. In most of eastern Mainland and in the eastern and northern isles the passage of ice has smoothed out the original relief, giving rise to rounded hills on which many of the original irregularities are now covered by a skin of boulder clay. U-shaped, glacially overdeepened valleys are rare. Possible examples occur at Dales Voe and Colla Firth in Delting and at Quarff, 8 km SW of Lerwick, where a major east–west trending valley cuts through the Clift Hills range, but the glacial origin of these is still in doubt. In Northmaven, Lunnasting, Muckle Roe and the northern part of the Walls Peninsula the craggy topography shows the effects of glacial moulding which has given rise to roches moutonnées and less well-defined ridges with intervening ice-gouged depressions. The glacial scouring has produced numerous hollows, most of which

---

[1] It has not been possible to show all the localities referred to in this chapter on the accompanying maps. The reader is referred to the one-inch-to-one-mile O.S. maps and the one-inch-to-one-mile drift maps published by the Institute.

now form small inland lochs. The straits between peninsulas and islands, such as Ronas Voe and the Swarbacks Minn (90 m deep) between Vementry and Muckle Roe, may also have been deepened by ice scouring. It is also likely that the deep sea-basins close to Shetland, such as St Magnus Bay (160 m deep) and the basin between Whalsay and Yell (145 m) may have resulted from glacial overdeepening, though Flinn (1970c) has suggested that these may represent earlier meteor impact craters which were cleared out by ice during the Pleistocene.

Interglacial Deposits

Deposits known to be older than Devensian have been recorded at only two localities in Shetland. On the west coast of Fugla Ness [312 913], North Roe, a bed of peat, up to 1·4 m thick, occurs between two layers of boulder clay (see Chapelhow 1965) and at Sel Ayre [177 541] on the west coast of the Walls Peninsula a 7·5 m thick deposit of bedded sand and gravel with up to 45 cm of peat near its base, and 3 m of boulder clay above it, occupies a preglacial or interglacial valley. Pollen analyses of the Fugla Ness (Birks and Ransome 1969) and Sel Ayre peats suggest that the deposits were formed during the Hoxnian interglacial stage, but, as there are no other known interglacial deposits within several hundred kilometres, any correlation must be tentative. Radiocarbon dates from Fugla Ness have given ages ranging from 35 000 to 40 000 BP (Page 1972), and a similar age was obtained from Sel Ayre (Mykura and Phemister 1976).

Devensian Glaciation

Deposits

The Shetland Islands are partly covered by a generally thin, irregular layer of stony till, with a matrix and pebble content which varies according to the character of the underlying bed rock and the type of rock over which the ice sheet has passed. Excellent temporary sections are seen from time to time in freshly opened roadside quarries and road cuttings but good permanent coast and stream sections are rare, the best being found on the shores of a few open voes and sounds, in drift-filled gullies and at the heads of some geos. Ice-transported erratics, some up to several metres in diameter, are common in many parts of Shetland. Mounds and ridges of morainic material, which are so common in the Scottish Highlands, are rare and generally small. Probably the best moraine belt extends for over 1·5 km across the centre of Papa Stour; smaller patches of hummocky moraine occur throughout central and southern Shetland.

Direction of Ice Movement

The pebble content of the boulder clay and moraine and the composition of the erratics have been used in conjunction with striated rock faces, chatter markings, roches moutonnées and other glacially moulded features to determine the directions and phases of ice movement (Fig. 28). Data from some areas such as Yell are still very scarce and only a few areas have been investigated in detail. The interpretations put forward in this chapter must therefore be regarded as tentative.

Maximum stage of glaciation. Peach and Horne's concept of an uninter-
rupted westward ice flow across Shetland during the maximum stage of the
last glaciation has not been substantiated, except perhaps in the extreme
north and south of the island group. In Unst boulders and pebbles of serpentine of the type cropping out in the east and centre of the island are present
on the high western Valla Field ridge and along the west coast to the south of
Wood Wick, indicating ice flow from the east-south-east. In North Roe
(north Mainland) a boulder clay exposed along the east coast, which has been
linked with either north-north-east trending or north-west trending striae,
contains pebbles which most probably come from north-easterly and south-
east to easterly directions. In South Mainland there is evidence that south of
the latitude of Quarff ice moved westward from the North Sea across the
high backbone of the island. Blocks of Old Red Sandstone conglomerate of
the type cropping out along the east coast, for example, were recorded by
Peach and Horne along the west shore near Wester Quarff and occur all along
the hilltops extending south from Quarff. Erratics of both Clift Hills phyllite
and Old Red Sandstone are common on St Ninian's Isle. It is also probable
that ice coming from the north-east covered part of Bressay, as in the northern
half of the island all the ice-transported pebbles and boulders consist of Old
Red Sandstone sediment and there are no stones of metamorphic rock of the
type found on Mainland.

In the central part of Shetland the evidence points to the existence of a
local ice sheet during the maximum stage of the Devensian. West of the central
backbone of Mainland the overall ice movement was to the west. The actual
direction was north-westward in North Roe, westward to west-north-west-
ward in Northmaven and Muckle Roe, almost radially outward from the
Walls Peninsula, and south-westward in the Scalloway–Burra Isle area.
There is an abundance of indicator stones throughout the area which confirm
this direction of movement, but there are no stones which could with
certainty have been derived from a more easterly source than the main
Scallafield ridge of central Mainland. In eastern Mainland the prevailing
orientation of striae swings from east-south-east and east–west in northern
Tingwall, Nesting, Laxo and on the south coast of Whalsay, to north-east
and even north-north-east in Lunnasting, central and north Whalsey and the
Out Skerries. In an unpublished report, now in an Institute open file,
Robertson (1935) has shown that, although the orientation of these striae is
in apparent accord with those farther west, the stones in the till and the
erratics invariably indicate an eastward or north-eastward direction of ice
transport. This contradicts the conclusions of Peach and Horne, who
visualised a south-westward ice movement in these areas. Rocks on which
Robertson's evidence is based include the migmatite and granite of the Colla
Firth permeation belt (Plate IV), boulders of which abound on the peninsulas
between Dales Voe and Wadbister Voe and in the South Nesting Peninsula,
and the Stava Ness Granite (p. 42) which provides many boulders in Whalsay,
and rarer boulders as far east as the Out Skerries. Boulder-trails are also
provided by smaller outcrops of distinctive rocks, such as the small bosses
of serpentine in Lunning and west Whalsay, blocks of which can be found
for some distance to the north-east. Farther north on the island of Fetlar,
Phemister has recorded many blocks of metamorphic rocks from the Lamb
Hoga Peninsula resting on the serpentines occupying the centre of the island,

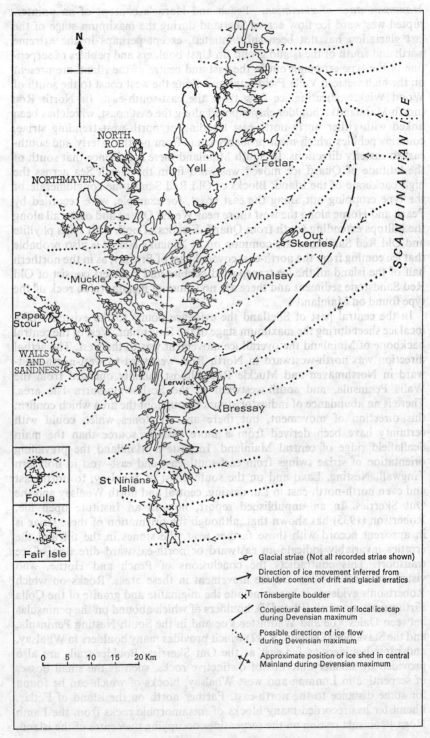

N

Unst

NORTH
ROE

Yell

Fetlar

NORTHMAVEN

Out
Skerries

DELTING

Muckle
Roe

Whalsay

Papa
Stour

WALLS
AND
SANDNESS

Lerwick

Bressay

SCANDINAVIAN ICE

Foula

Fair Isle

St Ninians
Isle

⊸ Glacial striae (Not all recorded striae shown)

→ Direction of ice movement inferred from boulder content of drift and glacial erratics

×T Tönsbergite boulder

— — Conjectural eastern limit of local ice cap during Devensian maximum

⋯⋯ Possible direction of ice flow during Devensian maximum

⤲ Approximate position of ice shed in central Mainland during Devensian maximum

0 5 10 15 20 Km

FIG. 28. *Directions of ice movement during the Devensian (Weichselian) in Shetland*

and abundant serpentines in the till overlying the Funzie conglomerate in the east. As no boulders of the staurolite-schist from the main Scallafield Ridge of central Mainland or of rocks cropping out even further west have been found in the eastern areas, Robertson concluded that central Shetland was covered by a local ice cap which had its ice-shed in the vicinity of Weisdale and Pettadale, whence it moved out both westward and eastward. The east-ward-flowing ice encountered resistance, presumably by Scandinavian ice, a short distance east of the coast of Mainland and then turned north-eastward past Whalsay and the Out Skerries.

A similar conclusion was reached independently by Flinn (1964, 1967b), who stated that, though eastern ice was carried westward across the watershed in the extreme north and the south of Shetland, in east-central Shetland a strong stream flowed eastward and then swung north as it extended farther east. The ice crossing Unst from the east may have been either deflected local ice or Norwegian ice. Hoppe (1970) however, though not disputing the facts enumerated above, has reverted to Peach and Horne's concept that Shetland was overridden by the Scandinavian ice sheet during the maximum stage of the Würm (i.e. Devensian) Period. He based this conclusion on the fact that there are striae on Bressay and on Weisdale Hill which indicate an early ice movement from the north-east and south-east respectively. It must be pointed out, however, that Weisdale Hill lies to the west of Robertson's postulated ice-shed and that the striae on Bressay could have been formed by ice which did not override Shetland Mainland. The present writer concludes that, though eastern ice may well have overridden Shetland at one period, the evidence for this has been obliterated by later ice, and that all the available evidence can be reconciled with Robertson's and Flinn's concept that during the Devensian maximum a local ice cap covered central Shetland and deflected the Scandinavian ice sheet around it.

Later ice cap. The weakening of the Scandinavian ice sheet and its subsequent break-up over the North Sea appear to have resulted in first a release of pressure on the locally generated ice and then its complete isolation. This enabled the local ice cap to expand and brought about an outward flow of ice throughout Shetland. The local ice eventually covered the part of southern Shetland which had earlier been overridden by eastern ice. Good evidence for an eastward ice flow in south-east Mainland is found in the Helli Ness Peninsula, where there are several large erratics of 'schist' and vein quartz derived from the Clift Hills and where the till contains many boulders of Cunningsburgh Spilite. Other evidence for eastward ice movement is found at Scousburgh, where large blocks of the basal Old Red Sandstone breccia occur up to 100 m E of the breccia outcrop, in the No Ness Peninsula near Sandwick where the drift contains both quartz and schist pebbles, and on the east shore of the Bay of Quendale near Sumburgh Airfield as well as on Sumburgh Head, where the till contains boulders and debris of Spiggie Granite. In the north-east part of North Roe Chapelhow has recognised a second till which contains boulders whose source lies to the south and west and which occupies much the same ground as the earlier till derived from the east (p. 107). In west Shetland it has not been possible to separate the deposits and striae of the later ice from those formed during the glacial maximum. There was here probably neither a time break nor a major directional change in the ice flow, though minor changes in direction may have taken place as the confining pressure exerted by the eastern ice diminished. As the ice broke up in the deeper bays around Shetland ice from the land would tend to flow directly into the bays, producing the ice flow at right angles to the present coast line.

It is not certain if at this period Unst and northern Yell were covered by the local ice sheet, as the evidence for eastward ice movement in Unst is weak. Nor is it certain that Bressay was covered by ice coming from Mainland. Peach and Horne have suggested that the Mainland ice was deflected south-eastward along the Sound of Bressay and that the Ward of Bressay may at this stage have supported its own ice cap.

Corrie Glaciation

Little is known about the stages in the melting and break-up of the Shetland ice cap. Its final remnants consisted of a number of small immature corrie glaciers and ice patches whose existence can be recognised by the presence of rudimentary corries and small morainic mounds and ridges (Charlesworth 1956). The largest of these are centred on Ronas Hill, and there are less well developed ones on the hills of central Mainland between Dales Voe to Weisdale Hill, on the eastern slope of the Clift Hills west of Cunningsburgh, and on the north and west slopes of Sandness Hill.

Glacial retreat features

Deposits and landforms produced by the melting of land ice are extremely rare in Shetland. Mounds, ridges, or spreads of glaciofluvial sand and gravel are virtually absent and glacial overflow channels are rare and poorly developed. In the Walls Peninsula there are a number of small isolated dry channels, which may have been formed by glacial meltwaters. The relative sea level around Shetland has risen considerably since the retreat of the ice, and, as the products and features of ice wastage in most upland areas are concentrated on the lower ground, it can be assumed that in Shetland they are now largely submerged.

Periglacial action on the higher plateaux has given rise to the blockfields (Felsenmeere) which cover Ronas Hill and the adjoining high plateaux and, to a less extent, Sandness Hill and the Ward of Culswick.

The best evidence relating to the date at which Shetland became free of ice has been obtained by the radiocarbon dating of the earliest late-glacial deposits taken from several Shetland lochs (Hoppe and others 1965, Hoppe 1970). These have given ^{14}C ages of about 10 000 years BP, which suggests that the greater part of Shetland became free of ice during the Alleröd interstadial or possibly slightly earlier. The small corrie glaciers may have lingered on considerably longer.

Changes in Sea Level

The presence of submerged peat beds, the lack of raised beaches and well-defined submerged shore platforms, the characteristic drowned river valleys now forming many of the Shetland voes, and the local traditions of submergence in historic times have all long been taken as evidence for the continuous submergence of the land since the Ice Age (*see* Finlay 1930). Submerged peat is common in the sheltered voes and sounds and has also been encountered in excavations in the harbours at Lerwick, Scalloway, Bressay, Graven (Sullom Voe) and Symbister (Whalsay). At Lerwick harbour peat was found at a depth of 6·4 m below high-water mark and at Symbister at between 8·6 and 8·9 m. The latter gave ^{14}C dates which range from 5455 BP to 6970 BP, indicating that some 5500 years ago sea level must have been *at least* 9 m lower than at present (Hoppe 1965). The subsequent rise in sea level may be partly accounted for by the world-wide eustatic rise in sea level which may have been as much as 6 m, and partly by the isostatic submergence of the land. It is probable that the isostatic depression of the Scottish mainland and Scandinavia during the glacial maxima was accompanied by uplift along the marginal zone of the ice sheets, which includes the Shetland area, and that the post-glacial submergence of Shetland is the result of recovery from this peripheral uplift (Flinn 1964).

Flinn (1964) has recognised submerged platforms around Shetland at the following levels (below OD): (1) 45 Fm (82 m), (2) 25 Fm (45 m), (3) 13 Fm (24 m), and possibly also (4) 5 Fm (9 m). Flinn believes that these shelves are not drowned post-glacial wave-cut platforms, but may represent remnants of earlier erosion surfaces. He suggests that in post-glacial times there have been no major pauses in the rise of sea level which would have permitted the cutting of wave-cut platforms.

Post-Glacial Deposits

As there are no major rivers, freshwater *alluvium* occupies only a very small proportion of the Shetland land surface. *Blown sand* covers a considerable area around the Bay of Quendale in south Mainland and somewhat smaller areas at Scousburgh, Bigton, Meil (Burra Isle), Melby and Papa Stour. There are also small areas of blown sand at West Wick and Inna Ness on the island of Yell and on Balta (Unst). Storm beaches are a feature of the exposed coasts. They are particularly well developed along the cliffs of Stenness and Esha Ness (Northmaven) where the Old Red Sandstone lavas and tuffs are readily broken up into blocks by wave action. At the Grind of the Navir, for instance, there is an arcuate accumulation of blocks on a waveswept platform at 15 m above sea level. Fine storm beaches are also seen along the cliff-tops on the south coast of the Out Skerries and, at a lower level, at Fugla Ness, near Hamnavoe on the Burra Isle.

Peat. Peat, mainly of the blanket bog type, covers a considerable part of Shetland and forms an almost unbroken cover over Yell, western Unst and large parts of central and western Mainland. It is virtually absent from the serpentine and greenstone areas of Unst and Fetlar and occupies only small patches in the rugged terrain floored by the diorite and granophyre of Northmaven and North Roe, the metamorphic rocks of the Walls Peninsula and the metamorphic rocks of the Scalloway area and the Burra Isle. On some islands, such as Papa Stour, all the peat has been removed for fuel.

The blanket bog does not normally exceed 1·5 m thickness, but in hollows and valleys up to 6 m of bog peat have been recorded. On Yell the peat reaches a thickness of 3 m, even on steep slopes. The stratigraphy and vegetational history of the Shetland peat was studied by Lewis (1907, 1911) who established the following sequence of vegetational zones:

B. *Upper unstratified peat.* (This has a greater extent than the lower peat, and rests in many places directly on weathered boulder clay.)

A. *Lower stratified peat*—consisting of:

    5. Second Arctic Bed

    4. Lower Peat Bog (up to 2 m+ thick)

    3. Forest Bed

    2. First Arctic Bed

    1. Basal peat (very local in extent), with aquatic plants.

These zones have not, as yet, been correlated with the pollen zones established in the peats of the Scottish mainland. According to Lewis the peat of the Forest Bed extends right up to the western seaboard of the Walls Peninsula, which indicates that the climatic regime at the time of its formation was very different from the present climate with its prevalent strong westerly winds.

Foula

Foula, with its relatively mountainous topography, was probably never completely overridden by ice from the east. Striated pavements with east to east-south-east trending scratches are seen on the north and east coasts of the island. The boulder clay exposed along these shores contains boulders of sandstone similar to that cropping out on the island, as well as metamorphic and igneous rocks. Erratics of epidotic granite, like the Spiggie Granite, are common. The island has three corries which face north-east to east and a U-shaped valley, the Daal, which may originally have terminated in a corrie just west of the present island. These features were formed by local ice which was probably already established at an early stage of the main Devensian glaciation and remained until the period of the Corrie Glaciation on Shetland Mainland. Ice flowing eastward from these corries was deflected by ice moving westward from Shetland Mainland both to the north and the south along the depression extending north and south from Ham. When the pressure from the eastern ice diminished local ice may have extended from the corries eastward beyond the limits of the island, thus accounting for the presence of sandstone boulders in some of the east coast till. At a later stage ice did not extend far beyond the higher corries, one of which has a fine arcuate terminal moraine at its mouth.

Fair Isle

Fair Isle was overriden by ice from the east-south-east, and this has obliterated all traces of any possible earlier glaciation. The westward ice movement appears to have been very powerful and has produced a strongly ice-moulded topography. This is particularly impressive in the southern half of the island (Plate XVI A), where the ice has gouged out depressions along east-south-east trending faults. The drift of Fair Isle is reddish brown in colour, has a relatively sandy matrix and contains a large number of small boulders and fragments of soft red sandstone. According to Flinn (1970a) both the till and the sandstone clasts contain millet-seed sand grains and he has suggested that the sediment from which these are derived may be of Permo-Triassic age. Flinn has also found evidence for an ice flow in a south-easterly direction

A. Fair Isle. Landscape moulded by ice crossing island from east (right) to west (left)
   (Aerofilms Ltd.)

Plate XVI

B. Tombolo ('ayre') of sand linking St. Ninians Isle to Shetland Mainland
   (Department of the Environment)

across Fair Isle. There is no indication that Fair Isle has at any time had a local ice cap.

Orkney

As in Shetland, the first connected account of the glaciation of the Orkneys was produced by Peach and Horne (1880). They showed that the ice which streamed eastwards from eastern Sutherland was deflected north-westwards and west-north-westwards over Caithness and Orkney by the Scandinavian ice sheet, and that after the retreat of this south-eastern ice, local glaciers lingered for a time in the valleys of Hoy and the higher parts of Mainland. The work of the Geological Survey confirmed Peach and Horne's principal findings but established that the glacial history was more complex than they had envisaged.

Over the greater part of Orkney the passage of the ice sheets has led to a smoothing out of the pre-existing topography. In the flagstone country the hillsides before glaciation were probably terraced by small rocky escarpments with scree at their base. The passage of ice has in most areas filled the depressions between escarpments with boulder clay and removed any projecting rock, leaving behind only indistinct ledges. In a few areas however, such as the south side of Fitty Hill in Westray and the south side of Rousay, ice advancing along the hill-slope scoured out the debris between escarpments and emphasised the terracing. North-west Hoy, which is the only part of Orkney to have supported local glaciers, has fairly well-developed corries on the east and west slopes of Ward Hill and on the north-west slope of Cuilags. The two glens on either side of Ward Hill, which converge to form the wide valley at Rackwick, have been considerably modified by valley glaciers.

Devensian Glaciation

Deposits

The boulder clay of Orkney is largely confined to the low ground, and is exposed in many excellent coast sections, where it ranges in thickness from 3 to 10 m. It generally consists of red or purple sandy clay with abundant polished and striated boulders. In eastern and many northern exposures the matrix is composed largely of material derived from the 'marls' and sandstones of the Eday Beds. Traced westward across West Mainland and Westray the red colour is gradually replaced by shades of brown, yellow and grey as the proportion of comminuted rock derived from the Rousay and Stromness Flags increases. Most boulders and smaller stones in the till consist of local material. Clasts derived from outside Orkney include various types of granitic, felsitic and schistose rocks, quartzites and quartzose sandstones, dark limestones with plant remains, calcified fossil wood, chalk and chalk flints. In the northern isles and parts of Mainland much of the till contains broken, smoothed and striated marine shells. Erratic boulders are relatively rare, the most interesting being a large boulder found near Saville in the north of Sanday which measures $2 \text{ m} \times 1.8 \text{ m} \times 0.75 \text{ m}$ and consists of an unusual variety of biotite-oligoclase-gneiss (Heddle 1880, Flett 1898a). This 'Saville Boulder' is now generally considered to be of Scandinavian origin. Other possible Scandinavian erratics have been recorded on the east shore of Flotta.

These consist of laurdalite, a variety of nepheline-syenite known to occur *in situ* in the Oslo district of Norway (Saxton and Hopwood 1919). The distribution of boulders in the till leaves no doubt that the principal ice movement was in a north-westerly direction. Examples are the abundance in Westray of blocks of red and buff sandstone of the type exposed on Eday, the presence in north-west Shapinsay of blocks of slaggy basalt which crops out in the south-east corner of the island, and the abundance of blocks of metamorphic rocks from the Stromness and Yesnaby outcrops on the ground west of these inliers. Most other igneous and metamorphic erratics can be matched with rocks cropping out in Sutherland. The Mesozoic rocks and marine shells are most probably derived from the bed of the North Sea to the east of northern Scotland and Orkney.

Striae

Striae are less abundant and more poorly preserved than in Shetland. Their predominant direction is north-westerly but there are also some with northerly, westerly and south-westerly trends (Fig. 29). Some glaciated surfaces in Rousay indicate two or three directions of ice movement. These suggest that the south-west and east–west trends are earlier than the north-westerly and northerly ones (Wilson 1935). From this it has been inferred that in the earlier stages of the Devensian (Würm) glaciation Scandinavian ice travelled westwards across Orkney, being locally deflected south-westward by the mountains of Hoy. Later the pressure exerted by the eastern ice appears to have decreased and Scottish ice was able to push its way north-westwards across the islands. This relief of pressure may have occurred more than once, and there may even have been times when Orkney was partly or completely free of ice.

Later local ice

Orkney never supported a local ice cap of the type that once covered Shetland, and the only evidence for local glaciers is found in north-west Hoy. Here the corries and glaciated valleys (p. 113) contain, or are associated with, morainic mounds. Hummocky moraines also occur on the hillside south-east of Rackwick (Hoy) and in the valley of the Forse Burn (Hoy).

Glacial retreat features

Morainic mounds, whose origin has not been fully explained, are found in the valley leading west from Finstown (West Mainland) and in the northern part of Mainland, near Loch Harray and Evie. These may be the deposits left by lobes of ice which, during a stage in the deglaciation, re-advanced westward and south-westward from the ice-filled bays and straits up the valleys of West Mainland.

Fluvioglacial deposits have only been recorded in western Hoy, where spreads of sand with gravel lenses occupy the floor of the valleys north of Rackwick.

Changes in Sea Level

Evidence for the gradual submergence of Orkney in post-glacial times is afforded by the absence of raised beaches, the presence in many bays of peat beds below the high-water mark and the existence at sea level of lochs with

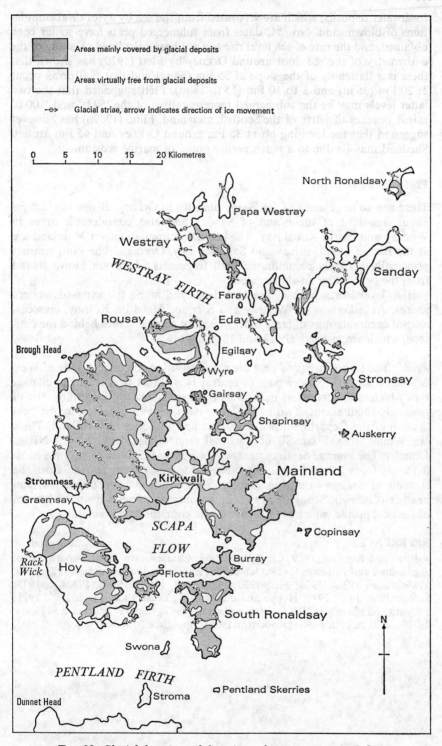

FIG. 29. *Glacial deposits and directions of ice movement in Orkney*

freshwater deposits, which are separated from the sea by ayres or accumulations of blown sand. No ^{14}C dates from submerged peats have so far been obtained, and the rate of sea level rise has not been calculated. A study of the bathymetry of the sea floor around Orkney by Flett (1920) has shown that there is a flattening of the slope at 35 Fm (64 m) and that platforms occur at 20 Fm (36 m) and 8 to 10 Fm (15 to 18 m). Flett suggested that the two latter levels may be the submerged representatives of the '50 ft' and '100 ft' raised beaches of parts of the Scottish mainland. Flinn (1969b) has however suggested that the levelling off at 35 Fm around Orkney and 45 Fm around Shetland may be due to a much earlier phase of marine erosion.

Post-Glacial Deposits

There are no large areas of *freshwater alluvium* in Orkney. *Blown sand* covers about one-third of the island of Sanday and also considerable areas in Westray and North Ronaldsay. The largest accumulations on Mainland are at the Bay of Skaill, Birsay and Sandside Bay, Deerness. The sand consists principally of finely comminuted shell fragments which are blown inland from the beaches by on-shore gales.

High-level storm beaches are best developed along the exposed western shores. At Aikerness on Westray and Sacquoy Head on Rousay, crescent-shaped accumulations of large blocks occur at some distance behind the cliff-tops, which are respectively 12 and 18 m high.

Peat. The largest areas of peat are those covering the eastern hills of West Mainland and the greater part of central Hoy. Most peat is of the blanket type which ranges from less than 50 cm to 1 m in thickness. Basin bogs are of relatively small extent. Two of the latter have been investigated by the Peat Section of the Department of Agriculture and Fisheries for Scotland. These are White Moss, 6 km SE of Kirkwall, and Glins Moss, 2·8 km NE of Dounby. The average depth of peat at White Moss is 3·4 m and at Glins Moss it ranges from 2·3 to 3·2 m. The bulk of the bog peat is formed from the remains of *Sphagnum* mosses together with cotton grass (*Eriophorum*) and heather (*Calluna*). Sedge (*Carex*) peat is commonly found near the bottom of the peat profile, which at Glins Moss also contains wood remains.

REFERENCES

  Birks and Ransom, 1969; Chapelhow 1965; Charlesworth 1956; Department of Agriculture and Fisheries for Scotland 1968; Engstrand 1967; Erdtman 1924, 1928, 1929; Finlay 1926a, 1930; Flett 1920; Flinn 1964, 1967b, 1969b, 1970a, c, 1973; Hoppe 1965, 1970, 1974; Hoppe and others 1965; Leask 1928; Lewis 1907, 1911; Mykura and Phemister 1976; Olsson and others 1967; Page 1972; Peach and Horne 1879b, 1880, *in* Tudor 1883; Robertson 1935 (unpublished); Wilson and others 1935.

12. ECONOMIC GEOLOGY

Ore Minerals

Shetland

Copper ores and associated minerals

At Sand Lodge mine, south Mainland [438 248], copper and iron ores have been intermittently worked from 1789 till the early 1920s. The ore occurs in a gangue of ankerite, siderite and calcite, which forms two north–south trending veins, 3 to 5 m thick and converging at depth. The most important ore minerals were recorded as *limonite, hematite, malachite* and some *chrysocolla* above the 30 m level and *chalcopyrite* and *siderite* below. It is estimated that the mine has yielded over 12 000 tons of ore. Another, smaller mine, was in operation around 1880 at Setter [435 256]. This yielded small quantitites of pure chalcopyrite. Adits have also been driven at Levenwick [409 217] and Hoswick [418 237], where both iron and copper ores might be expected. It is not known, however, what minerals, if any, were obtained. In the extreme south of Mainland, at Quendale [368 127], Wick of Shunni [35 15] and Garths Ness [364 113], adits have been driven for copper. Close to the last-named locality there is an ore body which forms a vertical lens up to 3·6 m wide and 30 m long, but consists largely of *pyrrhotite* with only small amounts of *pyrite* and *chalcopyrite*.

Small quantities of chalcopyrite, usually embedded in veins of ankerite and calcite, are to be found at the Bight of Vatsland [467 462]; at Muckle Hell [526 400] and Rules Ness [529 426], on the east coast of Bressay; in western Noss [532 405]; and at Aith Wick [445 297], Croo Taing [429 269], Sand Wick [437 237] and Levenwick Ness [417 214]. Chalcopyrite in quartz veins occurs at No Ness [445 212]. On Fair Isle the copper ores *digenite, chalcopyrite, bornite, covellite, tetrahedrite, cuprite* and *malachite*, together with *pyrite* and *goethite* have been recorded at Copper Geo [203 728] where they occur in a gangue of scapolite and calcite. In 1912 over 15 tons of this ore was quarried and some of it exported. Smaller quantities of copper ores, also associated with scapolite, have been found in the North and South Reevas [200 709] and in other localities in south-west Fair Isle. Minute quantities of chalcopyrite, associated with ankerite, occur at Duttfield in east Fair Isle [222 722].

Other non-ferrous ores

Galena has been recorded near Hamnavoe in southern Yell [500 799] and along the east coast of central Mainland at Vidlin Voe [481 666], Dury Voe, Lax Firth [446 486] and the Bight of Vatsland [467 458]. At Vidlin Voe the galena occurs within a 15 m wide band within the schists, which has a high concentration of pyrrhotite and some sphalerite and chalcopyrite. *Sphalerite*

117

(zinc blende) is also found on the shores of Dales Voe [456 455 and 455 450], and *gersdorffite*, a nickel-arsenic sulphide, occurs on the west shore of Aith Voe [422 291]. Some of these ores are set in a gangue of ankerite and calcite. At Mo Geo [443 283] in the south-west corner of the Helli Ness peninsula, at Blue Geo near Sand Lodge [438 247] and at Honga Ness in Fetlar [655 913] there are irregular dyke-like masses of ankeritic carbonate with specks of chalcopyrite. These are coloured bluish green by traces of *fuchsite*, a chrome-mica. *Chromite* is a common accessory constituent of the serpentines of Unst where it is locally sufficiently concentrated to be suitable for commercial exploitation. In the past it has been worked in quarries on the south slope of Nikka Vord [625 103] just north of Balta Sound, at Hagdale [638 102] and to a smaller extent near Hamaberg [596 037]. Between 1820 and 1944 some 50 000 tons of chromite ore were extracted in Unst, first for use in the chemical industry and, later, as a refractory and for the manufacture of chrome-magnesite bricks. In spite of extensive exploration in the early 1950s (*see* Rivington 1953) the deposits have not been worked in recent years. A number of relatively rare minerals are associated with the chromite of Unst. These include kämmererite (a chrome-chlorite), zaratite: $Ni_3(CO_3)(OH)_4 \cdot 4H_2O$, uvarovite: $Co_3Cr_2(SiO_4)_3$ and pentlandite: $(Fe, Ni)_9S_8$. Chromite also occurs in appreciable quantities in the serpentine of Hesta Ness, north-east Fetlar [663 927].

Iron Ores

Hematite, limonite, goethite and *siderite* are associated with the copper ores of Sand Lodge and Levenwick (p. 117). The Sand Lodge Mine has yielded a considerable tonnage of hematite, but this was of secondary importance to the copper ore. At Clothister Hill [342 729], Sullom, there is a lenticular ore body of *magnetite* which has a high degree of purity and an exceptionally low phosphorus content (60 to 67 per cent Fe and less than 0·006 per cent P). At the outcrop it has a north–south trend, a length of 53 m and an average width of 3 m. It has a steep westerly inclination and at a depth of 22 m it has an average width of 3·3 m, but thins out rapidly below that level. The estimated volume of the ore body is 4000 m^3, giving a possible 20 000 tons of ore. The magnetite was mined between 1954 and 1957 and used in the manufacture of heavy mud for a coal flotation process. Between 6000 and 10 000 tons were extracted at that time.

Orkney

It has been known for more than four centuries that lead and copper ores are to be found in Orkney, and many unsuccessful attempts have been made to work them economically.

Lead Ores

Galena is usually associated with the gangue minerals dolomite, calcite and strontianite, which fill veins and interstices in breccia and small faults. The largest lead-bearing deposit occurs on the coast near Warebeth [224 087], 2 km W of Stromness, where a vertical north-east trending vein was worked about 1775. Good exposures of the vein-breccia with small specimens of galena are still seen on the coast. Other old lead mines are situated at Manse

Bay [477 921] on the east coast of South Ronaldsay (worked in the latter half of the 18th century), on Graemsay (exact location not known, closed early in 17th century), in western Rousay [374 311] and in Sanday [702 422] (opened about 1880). Other occurrences of galena are at Mill Bay in Stronsay [656 267], on the shore of Mainland near Rennibister [398 131], at Burnside [259 104] 1·5 km N of Stromness, at the Bay of Navershaw [265 088] and in Walliwall Quarry [436 104] 1·5 km W of Kirkwall. There are also records and traditions of lead ore having been found in Deerness, St Andrews, Shapinsay, Fara and Yesnaby [22 15].

Copper ores

The primary copper ores met with in Orkney are *chalcopyrite, chalcocite* and some *native copper*. These occur in only small quantities. Secondary ores are *malachite, azurite* and *chrysocolla*. Attempts to work these ores have been made at Wha Taing [445 961] in the west of Burray (worked before 1774), and on Rousay [387 285]. Exposures of ore can still be seen at the old mill-stone quarry, Yesnaby [219 154].

Iron ores

Hematite, limonite and *goethite* have been worked at the Bay of Creekland [237 044] and at The Candle of the Sale [273 017] in northern Hoy. In both localities the ore forms thin veins lining joints and fractures.

Manganese ore

Manganese ore has been recorded and actually worked (prior to 1774) at Lead Geo on the western cliffs of Hoy [186 032], in a position some 60 m below the top of the 275 m high cliff.

Uranium ores

Disseminated uranium ore is present in the basal beds of the Stromness Flags in the Yesnaby and Stromness districts of West Mainland. Here dolomitic conglomerate abutting against the metamorphic inliers (p. 73) contains in places up to 300 ppm uranium and 500 ppm zinc, as well as some galena and baryte. The flags immediately above the conglomerates also contain some uranium.

Talc

Considerable quantities of talc are present along the margins of, and in places within, the serpentine blocks of Unst and Fetlar. In Unst talc has been wrought at Queyhouse Quarry [612 123], sited within the talc belt along the western margin of the Main Serpentine Block. This quarry has been in operation since 1945 and produces about 9000 tons per annum. The talc is used mainly in the manufacture of roofing felt. Talc is also present along the western margin of the Clibberswick Block, and the remains of ancient soap-stone quarries, which formed the basis for an ancient cooking pot industry, can still be seen at Clibberswick [652 121]. On Fetlar near-vertical bands of high quality talc cross the serpentine of Hesta Ness [662 925]. These were worked for a short time around 1914.

An extensive outcrop of talc-magnesite-schist, associated with serpentine

J

and metamorphosed basic igneous rocks, crops out in the vicinity of Cunningsburgh, Shetland south Mainland (Bain and others 1971). This contains several areas of good quality talc-magnesite rock. One of the largest and best exposed masses lies close to the Burn of Catpund, some 250 m W of the main road [425 271]. Here the rock appears to be almost homogeneous and covers an area of about 550 000 m². The remains of ancient (?Viking) quarries can be seen in this area. Talc and magnesite are present in roughly equal amounts in these deposits. They also contain up to 10 per cent magnetite, chromite and chlorite. Investigations have been carried out into methods of separating the talc from the other components. It should be possible to use the whole rock directly as a raw material for various industrial products, such as a filler for rubber, plastics etc.

Baryte

Baryte does not occur in commercial quantities in either Orkney or Shetland. Along the north-east coast and on parts of the south coast of Papa Stour it forms narrow veins emplaced along joints in the rhyolite cliffs. Small zones with thin baryte veins are also present in Lunning, at Muckle Hell on Bressay, and in western Noss. In Orkney baryte forms part of the gangue mineral of the galena veins at Warebeth, Rousay and Stronsay (pp. 118–119).

Kaolin

A vertical belt of soft kaolinitised metamorphic rock crops out along the course of the Tactigill [373 516] and Dale [372 482] burns, east and south of Tresta, Shetland Mainland. The belt varies in width from 9 to 30 m and contains up to 40 per cent kaolin. A much smaller deposit of relatively pure kaolin occurs close to the shore at Moo Wick [622 877] in the Lamb Hoga peninsula, Fetlar (May and Phemister 1968).

Semi-Precious and Lapidary Stones

Shetland

Though none of the minerals of Shetland can be classed as precious stones, there are many rocks and minerals suitable for cutting and polishing into ornaments, brooches etc. Some fine mineral specimens can also be obtained. *Serpentine* is abundant in Unst and Fetlar, but not all of it is of gemstone quality. On Mainland serpentine of good colour has been found on the north shore of Colla Firth [357 844]. There is also a suitable outcrop on Vementry Island near the north-west corner of Maa Loch [297 604]. *Calcite* of a fine pink to reddish colour forms thick irregular veins at the Nabb of Lerwick [479 403], on the north-east shore of Brei Wick [477 405] and the west coast of the Ness of Beosetter, Bressay [492 443]. Dyke-like masses of carbonate rock (mainly ankerite) coloured green by an admixture of chrome minerals crop out on the shore at Aith Voe [443 283], Sand Lodge [439 248] and eastern Fetlar [656 913]. *Magnetite* octahedra up to 1 cm in diameter, set in chlorite, crop out in a bay in the north-east of North Roe [386 936]; smaller octahedra occur in eastern Fetlar [670 914]. *Scapolite*, a pale bluish to purplish white mineral, which takes a good polish, forms a thick vein at Skelda Ness, at the

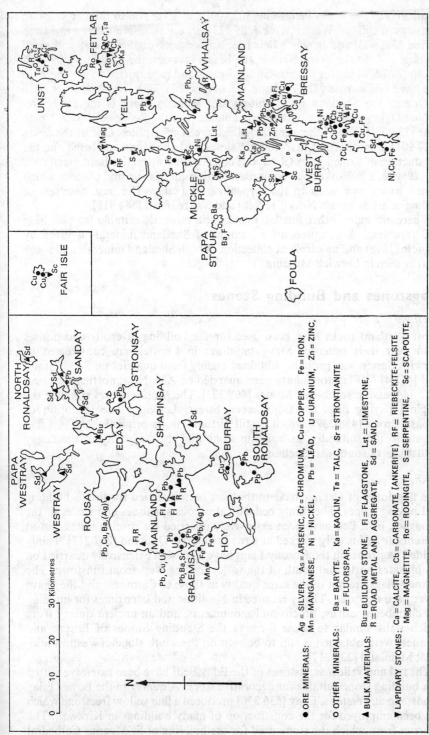

FIG. 30. *Location of main economic deposits (etc.) in Shetland and Orkney*

south end of the Walls Peninsula [303 404] and gives rise to thinner veins on the eastern cliffs of Wester Wick [285 423], in the roadside quarry at Mavis Grind [341 682] and in south Fair Isle. Rather poor quality *agates* and small amounts of *fluorite* are present in the basalt lavas on the west [149 609] and south [176 596] shores of Papa Stour. Agates and rare amethysts are found in the lavas of Stenness [214 772] and Esha Ness [203 780].

One of the Shetland rocks suitable for lapidary work is the *riebeckite-felsite*. Dykes of this are particularly abundant at the Beorgs of Uyea, North Roe [33 90]. Many of them contain bluish green spherulites. One of the dykes [327 901] was quarried in Neolithic times and the stone was used for the production of stone knives (Ritchie 1968). Another attractive hard green rock is *rodingite*, a jade-like rock composed of diopside, garnet and chlorite which forms lenses and veins in the serpentine of Fetlar. The best outcrops of rodingite are at Tressa Ness [617 948] and Swart Houll [644 918].

There are many other Shetland rocks and minerals suitable for polishing and tumbling. A comprehensive account of Shetland minerals is given by Heddle (1878) and an excellent collection of both Shetland minerals and rocks is to be seen in Lerwick Museum.

Flagstones and Building Stones

Shetland

Most Shetland rocks have been used for the building of crofts and houses sited near their outcrop. Many buildings in Lerwick are constructed of Lerwick Sandstone, probably obtained mainly from quarries on South Stony Hill [461 417]. Flagstones have been quarried on Aith Ness, northern Bressay [512 444] and in south-east Mousa [465 233]. The Mousa quarry supplied the original paving flags for Commercial Street, Lerwick. The condition of Mousa Broch [458 237], which is still virtually in its original state, is a fine tribute to the suitability as building stone and the durability of the flaggy sandstones of south-east Shetland.

Orkney

The flagstones of Orkney have in the past been quarried for use as building and paving stones. The thinly bedded calcareous siltstones and shales of the Sandwick Fish Bed have been extensively quarried for roofing slates, but at present the bed is only worked at Cruady Quarry, Quoyloo [247 217]. Thinly bedded flagstones in the Rousay Flags have also been extensively quarried on the hillsides north and south of the valley leading west from Finstown. The Orkney Flags are not of the same quality as those of Caithness and they have never been exported. They were used in Neolithic and later times for building the chambered cairns, brochs and monuments, and an ancient quarry from which slabs similar to those forming the Standing Stones of Brogar and Stenness were obtained is still to be seen on the south slope of Vestra Fiold, West Mainland [239 217].

The red and yellow sandstones of the Eday Beds have been extensively used as a building stone suitable for decorative work. A quarry in the Lower Eday Sandstone at Fersness, Eday [536 336] produced a fine yellow freestone which has been employed for the construction of many buildings in Kirkwall. The quarries from which the stone used for the building of St Magnus Cathedral was obtained were probably sited on the Head of Holland, 4 km ENE of Kirkwall. Much of the Upper Old Red Sandstone of Hoy appears to be a good quality building stone, but has only been used locally.

Road Metal and Aggregate

Shetland

Most of the igneous and many metamorphic rocks of Shetland have in the past been used for road metal and aggregate. At present most of the material is obtained from quarries at Scord, Scalloway [412 400] (phyllitic schist); Sullom ('granite'); Houlland, South Yell [50 80] ('gneiss'); Setter, Unst [636 115] (serpentine) and Whalsay [546 623] (gneiss). In the past beach gravels have also been locally used as a source for concrete aggregate.

Orkney

The calcareous flagstones of the Rousay and Stromness Flags are at present extensively used for road metal and aggregate. The principal quarries are at Cursiter [376 125], Walliwall [436 104] and Quoyloo [247 217] on Mainland, and there are smaller quarries on most of the other islands. The granite-gneiss of Brinkie's Brae, Stromness [252 096] has been quarried on a small scale in the past and should be suitable for both road metal and high quality aggregate. The basic dykes of Orkney are generally too thin for quarrying.

Limestone

Shetland

The thick crystalline limestones of east Mainland, Unst and the Out Skerries are all suitable for the production of lime for building and agricultural purposes. At present, however, the only quarry in operation is that working the Girlsta Limestone at Girlsta [429 505], 10 km N of Lerwick. The limestone contains nearly 44 per cent CaO and 5 per cent MgO. It has up to 10 per cent of quartz as well as accessory chlorite and muscovite (Muir and others 1956). The limestone is used for the production of agricultural lime. The lime workings at Girlsta and Ukensetter have been described by O'Dell (1939).

Orkney

Some of the calcareous, fish-bearing, shales in the Rousay Flags have a high lime content. Though they are less than 1 m thick they have been quarried in the past and used for the production of building lime. Examples of such quarries are to be found at Gallowhall in Evie [374 240] and Skel Wick in Westray [493 453].

Sand and Gravel

Both Orkney and Shetland are devoid of fluvioglacial sand and gravel, and there is very little river alluvium. The principal source of sand in the islands is from the accumulations of blown sand and from the sea beaches. In Orkney the blown sand is composed largely of comminuted shells and contains between

60 and 90 per cent CaCO₃. The principal deposits on Mainland are at Sandside Bay, Skaill, Deerness [589 067] where they are being extensively worked at present, as well as at Bay of Skaill [235 195], Birsay [248 270] and Aikerness [380 264]. Considerably larger deposits of blown sand cover nearly one third of Sanday and large parts of North Ronaldsay, Westray and Eday. Small areas of silica-sand consisting essentially of quartz grains have been recorded on beaches at Westray [479 462 and 472 447], Eday [564 343], Stronsay (Mill Bay) and Hoy [205 986]. The largest area of blown sand in Shetland is found at Quendale [380 130], but all the sand at present used commercially comes from sea beaches such as those at St Ninian's Isle and Gulber Wick.

Peat

Peat (p. 111) is extensively dug as a domestic fuel in both Shetland and Orkney, but there has as yet been no successful exploitation on a commercial scale. Peat briquettes were manufactured at one time at the Loch of Brindister [431 368], 5·5 km SW of Lerwick, but the venture was not financially viable. The greatest resources of peat are on the Island of Yell, which contains an estimated 200 million tons of raw peat or 16 million tons of peat solids. Considerable reserves are also available in parts of Shetland Mainland.

REFERENCES

Bain and others 1971; Department of Agriculture and Fisheries for Scotland 1968; Dron 1908; Gallagher and others 1971; Groves 1952; Heddle 1878, 1901; Hitchen 1929; Leask 1928; Macgregor and others 1920 (pp. 216–20); May and Phemister 1968; Muir and others 1956; Mykura 1972a, c; Mykura and Young 1969; Mykura and Phemister 1976; O'Dell 1939; Phemister 1964; Phemister and others 1950; Phillips 1927; Ritchie 1968; Rivington 1953; Robertson and others 1949; Sandison 1948; Scottish Council (Development and Industry) 1954a, b; Strahan and others 1916; Wilson 1921 (pp. 115–19, 147–51); Wilson and others 1935; Wilson and Phemister 1946.

13. GEOLOGICAL SURVEY MAPS OF ORKNEY AND SHETLAND

(a) **On the scale of 1 inch to 4 miles (1/253440).** Colour-printed.

Sheet 1 and 2 (combined)   Shetland.
Sheet 3   Orkney.

(b) **On the scale of 1 inch to 1 mile (1/63 360).** Colour-printed.

These sheets show both solid and drift deposits. Those marked * are printed in separate 'solid' and 'drift' editions. 117 (Hoy), 118 (Copinsay), 119 (Kirkwall), 120 (Stronsay), 121 (Westray), 122 (Sanday); Southern Shetland* (in preparation), Western Shetland*, Central Shetland* (in preparation), Northern Shetland*.

(c) **On the scale of 6 inches to 1 mile (1/10 560)**

The whole of Orkney and the greater part of Shetland is covered by geological maps on the six-inch scale. The remaining maps are in preparation. None of these are published, but they may be consulted at the Institute of Geological Sciences, Murchison House, West Mains Road, Edinburgh. Photostat copies (uncoloured) of these maps can be purchased at the discretion of the Assistant Director, Edinburgh.

14. BIBLIOGRAPHY

AGASSIZ, L. 1833–43. *Récherches sur les Poissons Fossiles*, Neuchatel.
—— 1834. On the Fossil Fishes of Scotland. *Rep. Br. Ass. Advmt. Sci.*, *4th Meeting, Edinburgh, Transactions of Sections*, 646–9.
AMIN, M. S. 1952. Metamorphic Differentiation of Talc-Magnesite-Chlorite Rocks in Shetland. *Geol. Mag.*, **89**, 97–105.
—— 1954. Notes on the Ultrabasic Body of Unst, Shetland Islands. *Geol. Mag.*, **91**, 399–406.
ANDERSON, F. W. 1950. Some reef-building calcareous algae from the Carboniferous rocks of Northern England and Southern Scotland. *Proc. Yorks. geol. Soc.*, **28**, 5–28.
APPLEBY, R. M. 1961. Continuation of the Great Glen Fault beyond the Moray Firth. *Nature, Lond.*, **191**, 1190.
RAIN, J. A., BRIGGS, D. A. and MAY, F. 1971. Geology and mineralogical appraisal of an extensive talc-magnesite deposit in the Shetlands. *Trans. Instn. Min. Metall.*, **80**, B, 77–84.
BALL, D. F. and GOODIER, R. 1974. Ronas Hill, Shetland: A preliminary account of its ground pattern features resulting from the action of frost and wind, *in The Natural Environment of Shetland*, edit. R. Goodier, 89–106.
BIRKS, H. J. B. and RANSOM, MAREE, E. 1969. An interglacial peat at Fugla Ness, Shetland. *New Phytol.*, **68**, 777–96.
BOTT, M. H. P. and WATTS, A. B. 1970. Deep Sedimentary Basins proved in the Shetland–Hebridean Continental Shelf and Margin. *Nature, Lond.*, **225**, 265–8.
BOUÉ, AMI. 1820. *Essai géologique sur L'Écosse*. Paris.
BRINDLEY, G. W. and VON KNORRING, O. 1954. A New Variety of Antigorite (Ortho-Antigorite) from Unst, Shetland Islands. *Am. Miner.*, **39**, 794–804.
CHALONER, W. G. 1972. Devonian plants from Fair Isle, Scotland. *Rev. Palaeobot. Palynol.*, **14**, 49–61.
CHAPELHOW, R. 1965. On Glaciation in North Roe, Shetland. *Geogrl. Jnl*, **131**, 60–70.
CHARLESON, M. M., ed. 1905. *Orcadian papers*; being selections from the *Proceedings of the Orkney Natural History Society*, from 1887 to 1904. Stromness.
CHARLESWORTH, J. K. 1956. The Late-glacial History of the Highlands and Islands of Scotland. *Trans. R. Soc. Edinb.*, **62**, 769–928.
CRAMPTON, C. B. and CARRUTHERS, R. G. 1914. The geology of Caithness. *Mem. geol. Surv. Gt Br*.
CURTIS, C. D. and BROWN, P. E. 1969. The metasomatic development of zoned ultrabasic bodies in Unst, Shetland. *Contr. Mineral. Petrol.*, **24**, 275–92.
——, BROWN, P. E. and SOMOGYI, V. A. 1969. A naturally occurring sodium vermiculite from Unst, Shetland. *Clay Miner.*, **8**, 15–19.
DEPARTMENT OF AGRICULTURE AND FISHERIES FOR SCOTLAND. 1968. *Scottish Peat Surveys, Vol. 4, Caithness, Shetland and Orkney*.
DONOVAN, R. N. and FOSTER, R. J. 1972. Subaqueous shrinkage cracks from the Caithness Flagstone Series (Middle Devonian) of northeast Scotland. *Jnl sedim. Petrol.*, **42**, 309–17.
—— FOSTER, R. J. and WESTOLL, T. S. 1974. A stratigraphical revision of the Old Red Sandstone of North-eastern Caithness. *Trans. R. Soc. Edinb.*, **69**, 167–201.

DOUBLE, I. S. 1939. Notes on some Recent Sediments from Unst, Shetland Isles. *Proc. Lpool. geol. Soc.*, **17**, 321–38.

DRON, R. W. 1908. Iron and copper mining in Shetland. *Trans. geol. Soc. Glasg.*, **13**, 165–9.

ENGSTRAND, L. G. 1967. Stockholm Natural Radiocarbon Measurements VII. *Radiocarbon*, **9**, 387–438.

ERDTMAN, G. 1924. Studies in the Micropalaeontology of the post-glacial Deposits in Northern Scotland and the Scotch Isles. *Jnl Linn. Soc.*, Botany, **46**, 449–504.

—— 1928. Studies in the Postarctic History of the forests of North-western Europe. Investigations in the British Isles. *Geol. For. Stockh. Forh.*, **50**, 123–92.

—— 1929. Aspects of the post-glacial history of British forests. *Jnl Ecol.*, **17**, 112–26.

FANNIN, N. G. T. 1969. Stromatolites from the Middle Old Red Sandstone of western Orkney. *Geol. Mag.*, **106**, 77–88.

—— 1970. The sedimentary environment of the Old Red Sandstone of western Orkney. *Ph.D. thesis, University of Reading* (unpublished).

FERNANDO, L. J. D. 1941. Petrology of Certain Felspathized Rocks from Herma Ness, Unst, Shetland Islands. *Proc. Geol. Ass.*, **52**, 110–30.

FINLAY, T. M 1926a. A Töngsbergite Boulder from the Boulder-clay of Shetland. *Trans. Edinb. geol. Soc.*, **12**, 180.

—— 1926b. The Old Red Sandstone of Shetland. Part I: South-eastern Area. *Trans. R. Soc. Edinb.*, **54**, 553–72.

—— 1930. The Old Red Sandstone of Shetland. Part II. North-western Area. *Trans. R. Soc. Edinb.*, **56**, 671–94.

FLEMING, J. 1811. Mineralogical Account of Papa Stour, one of the Zetland Islands. *Mem. Wernerian Nat. Hist. Soc.*, **1**, 162–75.

FLETT, J. S. 1897. On the discovery in Orkney of the John o'Groats Horizon of the Old Red Sandstone. *Proc. R. phys. Soc. Edinb.*, **13**, 255–7.

—— 1898a. On Scottish Rocks containing Orthite. *Geol. Mag.*, **5**, 388–92.

—— 1898b. The Old Red Sandstone of the Orkneys. *Trans. R. Soc. Edinb.*, **39**, 383–424.

—— 1900. The Trap Dykes of the Orkneys. *Trans R. Soc. Edinb.*, **39**, 865–905.

—— 1908. On the age of the Old Red Sandstone of Shetland. *Trans. R. Soc. Edinb.*, **46**, 313–9.

—— 1920. The Submarine Contours around the Orkneys. *Trans. Edinb. geol. Soc.*, **11**, 42–9.

FLINN, D. 1952. A Tectonic Analysis of the Muness Phyllite Block of Unst and Uyea, Shetland. *Geol. Mag.*, **89**, 263–72.

—— 1954. On the time relations between regional metamorphism and permeation in Delting, Shetland. *Q. Jnl geol. Soc. Lond.*, **110**, 177–99.

—— 1956. On the deformation of the Funzie conglomerate, Fetlar, Shetland. *Jnl Geol.*, **64**, 480–505.

—— 1958. On the nappe structure of North-East Shetland. *Q. Jnl geol. Soc. Lond.*, **114**, 107–36.

—— 1959. On certain geological similarities between north-east Shetland and the Jotunheim area of Norway. *Geol. Mag.*, **96**, 473–81.

—— 1961a. On Deformation at Thrust-Planes in Shetland and the Jotunheim area of Norway. *Geol. Mag.*, **98**, 245–56.

—— 1961b. Continuation of the Great Glen Fault beyond the Moray Firth. *Nature Lond.*, **191**, 589–91.

—— 1964. Coastal and Submarine Features Around the Shetland Islands. *Proc. geol. Ass.*, **75**, 321–39.

—— 1967a. The metamorphic rocks of the southern part of the Mainland of Shetland. *Geol. Jnl*, **5**, 251–90.

—— 1967b. Ice front in the North Sea. *Nature, Lond.*, **215**, 1151–4.

FLINN, D. 1969a. A geological Interpretation of the Aeromagnetic Maps of the Continental Shelf around Orkney and Shetland. *Geol. Jnl*, **6**, 279–92.

—— 1969b. On the development of coastal profiles in the north of Scotland, Orkney and Shetland. *Scott. Jnl Geol.*, **5**, 393–9.

—— 1970a. The Glacial Till of Fair Isle, Shetland. *Geol. Mag.*, **107**, 273–6.

—— 1970b. Some aspects of the geochemistry of the metamorphic rocks of Unst and Fetlar, Shetland. *Proc. geol. Ass.*, **81**, 509–27.

—— 1970c. Two possible submarine meteorite craters in Shetland. *Proc. geol. Soc. Lond.*, **1663**, 131–5.

—— 1973. The topography of the sea floor around Orkney and Shetland and in the northern North Sea. *Q. Jnl geol. Soc. Lond.*, **129**, 39–59.

—— 1974. The coastline of Shetland, *in The Natural Environment of Shetland*, edit. R. Goodier, 13–23.

——MAY, F., ROBERTS, J. L. and TREAGUS, J. E. 1972. A revision of the stratigraphic succession of the East Mainland of Shetland. *Scott. Jnl Geol.*, **8**, 335–43.

—— MILLER, J. A., EVANS, A. L. and PRINGLE, I. R. 1968. On the age of the sediments and contemporaneous volcanic rocks of western Shetland. *Scott. Jnl Geol.*, **4**, 10–19.

GALLAGHER, M. J., MICHIE, U. McL., SMITH, R. T. and HAYNES, L. 1971. New evidence of uranium mineralization in Scotland. *Trans. Inst. Min. Met.*, **80**, B150–173.

GARSON, M.S. and PLANT, J. A 1973. Alpine type utramafic rocks and episodic mountain building in the Scottish Highlands. *Nature Phys Sci.*, **242**, 34–38.

GEIKIE, A. 1877. The glacial geology of Orkney and Shetland. *Nature, Lond.*, **16**, 414–6.

—— 1879. On the Old Red Sandstone of Western Europe. *Trans. R. Soc. Edinb.*, **28**, 345–452.

—— 1897. *The Ancient Volcanoes of Great Britain*. 2 vols. London.

GILL, K. R. 1965. The petrology of the Brae Complex, Delting, Shetland. *Ph.D. Thesis, University of Cambridge* (unpublished).

GOODIER, R. (Editor). 1974. *The Natural Environment of Shetland*. Proceedings of Nature Conservancy Council Symposium.

—— (Editor). 1975 *The Natural Environment of Orkney*. Proceedings of Nature Conservancy Council Symposium.

—— and BALL, D. F. 1975. Ward Hill, Hoy, Orkney: Patterned Features and their origin, *in The natural Environment of Orkney*, edit. R. GOODIER, 47–56.

GROVES, A. W. 1952. Wartime Investigations into the Haematite and Manganese Ore Resources of Great Britain and Northern Ireland. *Monograph 20–703, Ministry of Supply Permanent Records of Research and Development*.

GUPPY, E. M. and SABINE, P. A. 1956. Chemical Analyses of Igneous Rocks, Metamorphic Rocks and Minerals: 1931–54. *Mem. geol. Surv. Gt Br.*

HEDDLE, M. F. 1878. *The County Geognosy and Mineralogy of Scotland, Orkney and Shetland*. Truro.

—— 1879a. The Geognosy and Mineralogy of Scotland, Orkney and Shetland. Island of Unst. *Mineralog. Mag.*, **2**, 12–35.

—— 1879b. Islands of Uya, Haaf Grunay, Fetlar, and Yell. *Mineralog. Mag.*, **2**, 106–33.

—— 1879c. The Mainland. *Mineralog. Mag.*, **2**, 155–90.

—— 1879d. A Brief Description of the Map of Shetland. *Mineralog. Mag.*, **2**, 253–5.

—— 1880. The Geognosy and Mineralogy of Scotland. Mainland [Shetland], Foula, Fair Isle. *Mineralog. Mag.*, **3**, 18–56.

—— 1901. *The Mineralogy of Scotland*. 2 vols. Edinburgh.

HIBBERT, S. 1819–20. Sketch of the Distribution of Rocks in Shetland. *Edinb. Phil. Jnl*, 1, 296–314; 2, 67–79, 224–42.

—— 1821. Discovery of Chromate of Iron in Shetland. *Trans. Soc. for the encouragement of Arts, Manufactures and Commerce*, 38, 23–6.

—— 1822. *A Description of the Shetland Islands*. Edinburgh.

HITCHEN, C. S. 1929. Unst and its chromite deposits. *Mining. Mag. Lond.*, 40, 18–24.

HOME, D. M. 1880. Valedictory address by President. *Trans. Edinb. geol. Soc.*, 3, 357–64.

HOOKER, J. D. 1853. Note on the Fossil Plants from the Shetlands. *Q. Jnl geol. Soc. Lond.*, 9, 49–50.

HOPPE, G. 1965. Submarine peat in the Shetland Islands. *Geogr. Annlr.*, 47A, 195–203.

—— 1970. The Würm ice sheets of northern and arctic Europe. *Acta. geogr. Univ. Lodz.*, 24, 205–15.

—— 1971. Nordvästeuropas inlandsisar under den sista istiden. *Särtryck ur Svensk Naturvetenskap*, 31–40.

—— 1974 The glacial history of the Shetland Islands. *Trans. Inst. Br. Geog.* Special publication No. 7 197–210.

—— SCHYTT, V. and STRÖMBERG, B. 1965. Från Fält och Forskning Naturgeografi vid Stockholms Universitet. *Särtryck ur Ymer.*, H, 3–4, 109–25.

INSTITUTE OF GEOLOGICAL SCIENCES. 1968. Aeromagnetic map of part of Great Britain and Northern Ireland. Sheet 16, Shetland sea area.

JAMESON, R. 1798. *An Outline of the Mineralogy of the Shetland Islands, and of the Island of Arran*. Edinburgh.

—— 1800. *Mineralogy of the Scottish Isles*. 2 vols. Edinburgh (Shetland, vol. 2, 185–224).

—— 1813. *Mineralogical travels through the Hebrides, Orkney and Shetland Islands, and mainland of Scotland*. 2 vols. Edinburgh.

KELLOCK, E. 1969. Alkaline basic igneous rocks in the Orkneys. *Scott. Jnl Geol.*, 5, 140–53.

KEY, R. M. 1972. A study of the metamorphic histories of the Saxa Vord, Valla Field and Lamb Hoga blocks of north-east Shetland. *Ph.D. thesis, University of Liverpool.* (unpublished).

LANKESTER, E. R. and TRAQUAIR, R. H. 1868–1914. The fishes of the Old Red Sandstone of Britain. *Mon. Pal. Soc.*

LAING, S. 1877. Glacial geology of Orkney and Shetland. *Nature, Lond.*, 16, 418–9.

LEASK, A. 1928. Shell Sand Deposits in Orkney. *Jnl Orkney agric. Discuss. Soc.*, 3, 57–58.

LEWIS, F. J. 1907. The Plant Remains in the Scottish Peat Mosses. III. The Scottish Highlands and The Shetland Islands. *Trans. R. Soc. Edinb.*, 46, 33–70.

—— 1911. The Plant Remains in the Scottish Peat Mosses. IV. The Scottish Highlands and Shetland, with an Appendix on Icelandic Peat Deposits. *Trans. R. Soc. Edinb.*, 47, 793–833.

LOW, G. 1879. *A tour through the North Isles and part of the Mainland of Orkney in 1774*. Kirkwall.

MACCULLOCH, J. 1821. The Elements of Practical Geology, London. 644–55.

—— 1831. A System of Geology. 2 vols. London, 1, 33, 70, 158, 208–10, 283; 2, 151, 158, 162, 169, 177, 193–7.

MACGREGOR, M., LEE, G. W., WILSON, G. V., with contributions by T. ROBERTSON and J. S. FLETT. 1920. The iron ores of Scotland. *Mem. geol. Surv. Gt Br. Min. Resources*, 11, 217–20.

—— and others. 1940. Synopsis of Mineral Resources of Scotland. *Mem. geol. Surv. Gt. Brit. Min. Resources*, 33.

McQUILLIN, R. 1968. Geophysical surveys in the Orkney Islands. *Geophys. Pap.* No. 4, *Inst. geol. Sci.*, 1–18.

—— and BROOKS, M. 1967. Geophysical surveys in the Shetland Islands. *Geophys. Pap.* No. 2, *Inst. geol. Sci.*, 1–22.

MATHER, A. S. and SMITH, J. S. 1974. The Beaches of Shetland. *Dept. Geog., University of Aberdeen.*

—— RITCHIE, W. and SMITH, J. S. 1974 Beaches of Orkney. *Dept. Geog., University of Aberdeen.*

—— 1975. An Introduction to the Morphology of the Orkney Coastline, *in The Natural Environment of Orkney*, edit. R. GOODIER, 10–18.

MAY, F. 1970. Movement, Metamorphism and Migmatization in the Scalloway Region of Shetland. *Bull. geol. Surv. Gt Br.*, No. 31, 205–26.

—— and PHEMISTER, J. 1968. Kaolin deposits in the Shetland Islands, U.K. *Rep. XXIII Int. Geol. Congr.*, Prague, 14, 23–9.

MILES, R. S. and WESTOLL, T. S. 1963. Two New Genera of Coccosteid Arthrodira from the Middle Old Red Sandstone of Scotland and their stratigraphical Distribution. *Trans. R. Soc. Edinb.*, 66, 179–210.

MILLER, H. 1849. *Footprints of the Creator*. Edinburgh.

—— 1858. *The cruise of the Betsey*. Edinburgh.

MILLER, J. A. and FLINN, D. 1966. A Survey of the Age Relations of Shetland Rocks. *Geol. Jnl*, 5, 95–116.

MUIR, A., HARDIE, H. G. M., MITCHELL, R. L. and PHEMISTER, J. 1956. The Limestones of Scotland: Chemical Analyses and Petrography. *Mem. geol. Surv. Gt Br. Min. Resources*, 37.

MUIR, R. O. and RIDGWAY, J. M. 1975. Sulphide Mineralisation of the Continental Devonian Sediments of Orkney (Scotland). *Mineral Deposita* (Berlin), 10, 205–215.

MURCHISON, R. I. 1853. Note of the Age and Relative Position of the Sandstone containing Fossil Plants at Lerwick in the Shetland Isles. *Q. Jnl geol. Soc. Lond.*, 9, 50–1.

—— 1859. On the Succession of the Older Rocks in the Northernmost Counties of Scotland: with some Observations on the Orkney and Shetland Islands. *Q. Jnl geol. Soc. Lond.*, 15, 353–418.

MYKURA, W. 1969. (for 1968). Geological Investigations on Fair Isle. *Fair Isle. Birds Obs. Rep.*, 19, 59–62.

—— 1970. Late- or post-Devonian vulcanicity in the Shetland Islands. *Proc. geol. Soc. Lond.*, No. 1663, 173–5.

—— 1972a. Tuffisitic breccias, tuffisites and associated carbonate-sulphide mineralization in south-east Shetland. *Bull. geol. Surv. Gt Br.*, No. 40, 51–82.

—— 1972b. The Old Red Sandstone sediments of Fair Isle, Shetland Islands. *Bull. geol. Surv. Gt Br.*, No. 41, 1–31.

—— 1972c. Igneous intrusions and mineralization in Fair Isle, Shetland Islands. *Bull. geol. Surv. Gt Br.*, No. 41, 33–53.

—— 1974. The Geological Basis of the Shetland Environment, *in The Natural Environment of Shetland*, edit. R. GOODIER, 1–12.

—— 1975a. The Geological Basis of the Orkney Environment, *in The Natural Environment of Orkney*, edit. R. GOODIER, 1–9.

—— 1975b. Possible large-scale sinisiral displacement along the Great Glen Fault in Scotland. *Geol. Mag.* 112, 91–93.

—— and PHEMISTER, J. 1976. The geology of western Shetland. *Mem. geol. Surv. Gt Br.*

—— and YOUNG, B. R. 1969. Sodic scapolite (dipyre) in the Shetland Islands. *Rep. Inst. geol. Sci.*, No. 69/4.

NICOL, J. 1844. *Guide to the Geology of Scotland*. Edinburgh. (The Zetland or Shetland Islands, pp. 243–246).

O'DELL, A. C. 1939. *The historical geography of the Shetland Islands.* Lerwick.

OLSSON, INGRID U., STENBERG, A. and GÖKSU, Y. 1967. Uppsala Natural Radio-carbon Measurements VI. *Radiocarbon*, 9, 454–70.

PAGE, N. R. 1972. On the age of the Hoxnian interglacial. *Geol. Jnl*, 8, 129–42.

PEACH, B. N. and HORNE, J. 1879a. The Old Red Sandstone of Shetland. *Proc. R. Phys. Soc. Edinb.*, 5, 80–7.

—— —— 1879b. The Glaciation of the Shetland Isles. *Q. Jnl geol. Soc. Lond.*, 35, 778–811.

—— —— 1880. The Glaciation of the Orkney Islands. *Q. Jnl geol. Soc. Lond.*, 36, 648–63.

—— —— 1884. The Old Red Volcanic Rocks of Shetland. *Trans. R. Soc. Edinb.*, 32, 359–88.

—— —— 1893. On the Occurrence of Shelly Boulder Clay in North Ronaldshay, Orkney. *Trans. Edinb. geol. Soc.*, 6, 309–13.

PEACH, C. W. 1865. Traces of Glacial Drift in the Shetland Islands. *Rep. Br. Ass. Advmt. Sci., 34th Meeting, Bath, 1864*, Transactions of Sections, 59–61.

—— 1879. A list of Fossil Plants, collected in Shetland, by Messrs. B. N. Peach and John Horne, of the Geological Survey, 1878. *Q. Jnl geol. Soc. Lond.*, 35, 811.

PHEMISTER, J. 1926. The Distribution of Scandinavian Boulders in Britain. *Geol. Mag.*, 63, 433–54.

—— 1964. Rodingitic Assemblages in Fetlar, Shetland Islands, Scotland. *Advancing Frontiers in Geology and Geophysics*, 279–95.

—— 1976. The Lunnister Metamorphic Rocks, Northmaven, Shetland. *Bull. geol. Surv. Gt Br.*, in press.

—— SABINE, P. A. and HARVEY, C. O. 1950. The riebeckite-bearing dikes of Shetland. *Mineralog. Mag.*, 29, 359–73.

PHILLIPS, F. C. 1926. Note on a Riebeckite-Bearing Rock from the Shetlands. *Geol. Mag.*, 63, 72–7.

—— 1927. The Serpentines and Associated Rocks and Minerals of the Shetland Islands. *Q. Jnl geol. Soc. Lond.*, 83, 622–51.

—— 1928. Petrographic Notes on Three Rock-types from the Shetland Islands. *Geol. Mag.*, 65, 500–7.

PLANT, J., SIMPSON P. R. and BOWLES, J. F. W. The origin and distribution of native metal, alloy, sulphide and arsenide phases in the Unst ophiolite assemblage, Shetland, *Trans. Inst. Min. Met.*, in press.

PRINGLE, I. R. 1970. The structural geology of the North Roe area of Shetland. *Geol. Jnl*, 7, 147–70.

READ, H. H. 1933. On quartz-kyanite rocks in Unst, Shetland Islands, and their bearing on metamorphic differentiation. *Mineralog. Mag.*, 23, 317–28.

—— 1934a. On the segregation of quartz-chlorite-pyrite masses in Shetland igneous rocks during dislocation-metamorphism, with a note on the occurrence of boudinage-structure. *Proc. Lpool. geol. Soc.*, 16, 128–38.

—— 1934b. On zoned associations of antigorite, talc, actinolite, chlorite, and biotite in Unst, Shetland Islands. *Mineralog. Mag.*, 23, 519–40.

—— 1934c. The Metamorphic Geology of Unst in the Shetland Islands. *Q. Jnl geol. Soc. Lond.*, 90, 637–88.

—— 1936a. The Metamorphic History of Unst, Shetland. *Proc. Geol. Ass.*, 47, 283–93.

—— 1936b. Summer field meeting, 1936, Unst, Shetland. *Proc. Geol. Ass.*, 47, 295–300.

—— 1937. Metamorphic Correlation in the Polymetamorphic Rocks of the Valla Field Block, Unst, Shetland Islands. *Trans. R. Soc. Edinb.*, 59, 195–221.

—— and DIXON, B. E. 1932. On stichtite from Cunningsburgh, Shetland Islands. *Mineralog. Mag.*, 23, 309–16.

RICHARDSON, J. B. 1965. Middle Old Red Sandstone spore assemblages from the Orcadian basin, north-east Scotland. *Palaeontology*, **7**, 559–605.

RIDGWAY, J. M. 1974. Sedimentology and Palaeogeography of the Eday Group, Middle Old Red Sandstone, Orkney. *Ph.D. thesis, University of London* (unpublished).

RITCHIE, P. R. 1968. The stone implement trade in third-millennium Scotland, *in* COLES, J. M. and SIMPSON, D. D. A. (editors). *Studies in Ancient Europe*, pp. 117–36, Leicester.

RIVINGTON, J. B. 1953. Recent Chromite Exploration in Shetland. *Min. Mag. Lond.*, **89**, 329–37.

ROBERTSON, T. 1935. The Glaciation of Aithsting, South Nesting, Whalsay and the Out Skerries. *Geological Survey Records*, (unpublished).

—— 1938. Observations on the direction of lineation in parts of Shetland. *Mem. Geol. Surv., Sum. Prog. for 1936*. Part 2, 75–8.

—— SIMPSON, J. B. and ANDERSON, J. G. C. 1949. The Limestones of Scotland. *Mem. Geol. Surv. Gt Br. Min. Resources*, **35**, 171–5.

SANDISON, C. G. D. 1948. *Historical account of the mineral workings in Unst, Shetland*. Scottish Council (Development and Industry), Paper No. 49(1) (unpublished).

SAXTON, W. I. and HOPWOOD, A. T. 1919. On a Scandinavian erratic from the Orkneys. *Geol. Mag.*, **56**, 273–4.

SCOTTISH COUNCIL (DEVELOPMENT AND INDUSTRY). 1954a. *Report of the Mineral Resources Panel on Serpentine and Olivine-rock in Scotland*.

—— —— —— —— 1954b. *Report of the Mineral Resources Panel on Talc in Scotland*.

SHIRREFF, J. 1817. *General View of the Agriculture of the Shetland Islands*. Edinburgh.

SINCLAIR, J. 1791–99. *The Statistical Account of Scotland*. 21 vols. Edinburgh.

SNELLING, N. J. 1957. A contribution to the mineralogy of chloritoid. *Mineralog. Mag.*, **31**, 469–75.

—— 1958. Further Data on the Petrology of the Saxavord Schists of Unst, Shetland Isles. *Geol. Mag.*, **95**, 50–6.

SPEED, J. 1666. England, Wales, Scotland and Ireland described.

STEAVENSON, A. G. 1928a. Some Geological Notes on Three Districts in Northern Scotland. *Trans. geol. Soc. Glasg.*, **18**, 193–233.

—— 1928b. The Geology of Stronsay Parish, Orkney. *Proc. Orkney Nat. Hist. Soc.*, 1–19.

STEPHEN, I. 1955. An occurrence of palygorskite in the Shetland Isles. *Mineralog. Mag.*, **30**, 471–80.

STRAHAN, A., FLETT, J. S. and DINHAM, C. G. 1916. Potash-felspar—phosphate of lime—alum shales—plumbago or graphite—molybdenite—chromite—talc and steatite (soapstone, soap-rock and potstone)—diatomite. *Mem. Geol. Surv. Gt Br. Min. Resources*, **5**.

SUMMARIES OF PROGRESS OF THE GEOLOGICAL SURVEY OF GREAT BRITAIN. 1930 (for 1929). *Mem. geol. Surv. Gt Br.*, 81–6.

—— 1931 (for 1930). *Mem. geol. Surv. Gt Br.*, 65–74.

—— 1932 (for 1931). *Mem. geol. Surv. Gt Br.*, 61–3, 79, 157–66.

—— 1933 (for 1932). *Mem. geol. Surv. Gt Br.*, 77–80, 95–7.

—— 1934 (for 1933). *Mem. geol. Surv. Gt Br.*, 70–6, 91–2.

—— 1935 (for 1934). *Mem. geol. Surv. Gt Br.*, 67–9, 82–3.

—— 1958 (for 1957). *Mem. geol. Surv. Gt Br.*, 42–3, 50.

—— 1959 (for 1958). *Mem. geol. Surv. Gt Br.*, 49–50, 53.

TRAILL, T. S. 1806. Observations, chiefly mineralogical, on the Shetland Islands, made in the course of a tour through these islands in 1803. *A Journal of Natural Philosophy, Chemistry and the Arts*. (*Editor Wm. Nicholson*). **15**, 353–67.

TRAQUAIR, R. H. 1908. On Fossil Fish Remains from the Old Red Sandstone of Shetland. *Trans. R. Soc. Edinb.*, 46, 321–9.

TUDOR, J. 1883. *The Orkneys and Shetland; Their Past and Present State*. London.

TUFNELL, H. 1853. Notice of the discovery of fossil plants in the Shetland Islands. *Q. Jnl geol. Soc. Lond.*, 9, 49.

WALKER, F. 1932. An albitite from Ve Skerries, Shetland Isles. *Mineralog. Mag.*, 23, 239–42.

WATSON, D. M. S. 1932. On three new species of fish from the Old Red Sandstone of Orkney and Shetland. *Mem. geol. Surv. Summ. Prog. for 1931*, Part 2, 157–66.

—— 1934. Report on Fossil Fish from Sandness, Shetland. *Mem. geol. Surv. Summ. Prog. for 1933*, Part 1, 74–6.

WATTS, A. B. 1971. Geophysical investigations on the continental shelf and slope north of Scotland. *Scott. Jnl Geol.*, 7, 189–218.

WESTOLL, T. S. 1937. The Old Red Sandstone Fishes of the North of Scotland, particularly of Orkney and Shetland. *Proc. Geol. Ass.*, 48, 13–45.

—— 1951. The Vertebrate-bearing Strata of Scotland. *Rep. XVIII Int. Geol. Congr.*, Part 11, Great Britain, 1948, 5–21.

WILSON, C. D. V. 1965. A marine magnetic survey near Lerwick, Shetland Islands. *Scott. Jnl Geol.*, 1, 225–30.

WILSON, G. V. 1921. The lead, zinc, copper and nickel ores of Scotland. *Mem. geol. Surv. Gt Br. Min. Resources. Gt Br.*, 17.

—— EDWARDS, W., KNOX, J., JONES, R. C. B. and STEPHENS, J. V. 1935. The Geology of the Orkneys. *Mem. geol. Surv. Gt Br*.

—— and KNOX, J. 1936a. The Geology of the Orkney and Shetland Islands. *Proc. Geol. Ass.*, 47, 270–82.

—— —— 1936b Orkney and Shetland Field Meeting. *Proc. Geol. Ass.*, 48, 61–76.

—— and PHEMISTER, J. 1946. Talc, other magnesium minerals and chromite associated with British serpentines. *Geol. Surv. Gt Br. Wartime Pamphlet*, No. 9.

15. APPENDIX I

Glossary of Some Orkney and Shetland Place Names

Aith—isthmus
-Ay, -a—island
Ayre—gravel beach or spit
Baa—flat submerged rock
Bak, bakka—steep bank, ridge
Bard, bord, barth—promontory, headland
Ber, berg, berri—rocky hill
Bjorgs, beorgs—long rocky ridge
-Bister, -buster, -vister—dwelling, farm
Bod, baa—site of a loch
Borg, burra, breck—slope
Brei—broad
Brett—steep
Brim—surf
Bring—breast
Bro—bridge
Brough, -burgh, Bur-—fort
Bu—farm, stock of cattle
Clett, klett (klettr)—low rock, seastack, cliff
Corbie—crow, raven
Croo—sheepfold, small walled enclosure
Dall, daal—dale, valley
Djub, djubi—deep
Drang, drong—pointed rock or stack
Erne—sea eagle
Evie—eddy
Ey, Eyn—island
Faer, Fair—sheep
Field, fjeld, fiold, felt, feal, fiel—hill
Fitful (Fitfugl)—web-footed bird
Flot (Flotta)—flat
For—patch of arable land
Ful, Fugl—bird
Galti—boar
Garth, gaard—enclosure
Geo (gja)—cleft, chasm
Gil—ravine
Gloup—throat, deep cut
Gos—meadow
Gra—grey
Grind—gate
Gron—green
Grut—gravel
Haaf—outmost
Hafr, Havr—goat
Ham—harbour
Hamar—rocky or rock-ledged hill
Hel, Hellia—flat rock, cave
Heog, hjog—hill or ridge with stony mounds
Hest, hestr—horse, stallion
Hevdi—headland
Ho, ha—high
Huga—pasture
Holm—small island
Houb—land-locked bay
Howe (haugr)—mound, burrow
Hope, hop—bay
Hov, hof—temple, mansion, fenced place
Hoy (hár)—high
Hund—dog
Hval, hwal—bay
Kame—ridge, crest
Kro—crow
Krokla, kroklin—mussel
Lamb—lamb
Lax—salmon
Lee—slope
Ler, leir—clay
Lir, liri—shearwater (bird)
Litla—small
Lun, lund (lundr)—grove
Mel, melr—sandbank
Minn—narrow sound
Mons—moor
Moul, Mul (melt)—rounded high promontory
Mur, mor—more
Nab—small promontory
Neep (gnípa)—steep hill
Nev, Nevi—nose
Ness (nes)—promontory
Noup, nip, nop, nup (gnupr)—high steep slope, generally to the sea
Oyce, ouse (os)—wide burn mouth, fresh-water loch held up by an ayre

Papa, papil—priest (Norse name for the Irish Culdee Priests)
Pund—enclosure, pound
Quoy (kvi)—cattle pen
Ram, ramm—raven
Riv, riva—gash, cleft
Ro, roe—red
Ron (Ronas)—stony stretch of hill
Röst, Roost—tide race
Scarf, Skarf—barren stony ground, shag
Sheen, shun, sjen, sjon—small lake
Setter, -ster (setr-)—summer homestead, out-pasture
Shun—small loch
Skaill, scall (skali)—large house
Skel—soft rock, shale
Skeld—strips of field
Skeir, Sker—skerry
Skord, scord—defile, hill-pass
Stav—stick
-Ster, -sta—homestead, farm
Stour—big
Strom—tide race
Sula—gannet

Sula—pillar
Swarback—black-backed gull
Swart—black
Taing—tongue
Teisti—black guillemot
Tind—spike, high peaked cliff
Ting—parliament, assembly
Toft, taft, topt—house site, plot of land
Tonga—tongue
Too—mound
Toun, Tun—home field, farmyard
Voe (vágr)—narrow bay
Va, waith—ford
Vat, vatn, van, watten—water, lake
Vin, vinya, winnya—pasture
Wall (vagr)—narrow bay (corruption of voe)
Ward, wart, vird, virt, vord—high look-out point on hill
Watten, vatn—water
Wick (vik)—bay, inlet
Note: The first syllables of many place names are derived from personal names.

REFERENCES

JAKOBSEN, J. 1936. *The Place Names of Shetland*. London and Copenhagen.
ORDNANCE SURVEY. *Place names on maps of Scotland and Wales*.

16. APPENDIX II

Guide to Geological Excursions

This section is intended for visitors to Orkney or Shetland who wish to see something of the geology of the islands and want some guidance as to good and readily accessible exposures. It is not possible to include a full excursion guide in this handbook and the following pages provide only the barest outline itineraries for 10 one-day excursions in Orkney, 11 on Shetland Mainland and a further 5 on the more accessible smaller Shetland islands. Some alternative excursions are also suggested. Excursions marked with * are recommended for those who can spend less time on the islands.

A. Orkney

1. *Stromness district*

   *(a) Shore section from The Ness [257 057] to Warebeth [234 087] and beyond, exposing conglomerates at base of Stromness Flags, typical rhythmic sequences in Stromness Flags with characteristic sedimentary structures and stromatolite mats and mounds. Sandwick Fish Bed cycle at Noust of Nethertown. Baryte-strontianite vein with galena at Warebeth.
   (b) Brinkies Brae [253 097] and Hillhead area west of Stromness. Granite and gneiss of Basement Complex.
   (c) Bay of Navershaw [266 099]. Porphyritic felsite overlain by felsite-conglomerate and flags.

2. *Yesnaby*

   *(a) Coast between Geo of Inganess [219 143] and Bor Wick [222 168]. Junction between metamorphic basement and Harra Ebb Formation; sediments of Harra Ebb Formation and angular unconformity at base of Stromness Flags; Harra Ebb cryptovents; Yesnaby Sandstone Formation with lower ?dune-bedded sequence around Stack of Yesnaby and upper fluvial sequence at Qui Ayre millstone quarry. Lower Stromness Flags sequence from Brough of Bigging northward with 'Horsetooth Stone' (digitate stromatolite) at [220 161].
   (b) *Quoyloo Quarry*, Hill of Cruday [246 216]. Sandwick Fish Bed. Best locality in Orkney for fish remains.

3. *Birsay and Evie*

   (a) *North-west coast of Mainland*, between Whitaloo Point [262 288] and Brough Head [233 287]. Upper Stromness Flags including 'Hoy Cycles' showing variations in rhythmic units. Fish beds. Sandwick Fish Bed Cycle at [247 283]. Good exposures of camptonite dykes and a tight monoclinal fold (ruck) in the flags at [259 288] (see Plate XI B). Lowish

136

tide preferable for most of this section, particularly the crossing to Brough of Birsay.

(b) *Evie district*, north shore of Mainland. Buckquoy [360 272]. Strata at junction of Upper Stromness Flags and Rousay Flags. Aiker Ness [384 268] Rousay Flags, fine broch.

4. *Deerness*

*(a) West coast of Newark Bay from [562 032] to [568 042]. Continuous sequence from Lower Eday Sandstone through Eday Flags to Middle Eday Sandstone, including teschenite intrusion and tuff beds at Muckle Castle [563 032]. (Low tide an advantage.)

*(b) Point of Ayre [593 039]. Basalt lavas and, at Greenigeo Taing [593 051], rhythmic units in Eday Flags, Lower Eday Sandstone.

*(c) East coast, at Tommy Tiffy [590 052]. Rousay—Eday Passage Beds, fish beds in Rousay Flags.

(d) Taracliff Bay, [554 034] to [557 033]. Fine section of Passage Beds and Lower Eday Sandstone.

5. *South Ronaldsay*

*(a) Barthwick [434 863] to The Kist [433 870]. Passage Beds between Rousay Flags and Lower Eday Sandstone. Vent of sediment-breccia and vent of monchiquitic agglomerate (shore, pebbles only accessible).

(b) Wind Wick, south shore [458 872] to Ossi Taing [462 866]. Eday Flags and Lower Eday Sandstone with fish beds.

(c) Sand Wick, west shore [430 892]. Typical red Middle Eday Sandstone.

*(d) Rumley Point (Croo Stone) [498 943]. Volcanic vent bounded and cut by dykes of monchiquite. Low tide essential.

6. *Hoy I*

(a) North coast between Bay of Creekland [237 047] and Middle Head [220 051]. Stromness Flags including Sandwick Fish Bed, Hoy Cycles and Upper Stromness Flags. Tuff beds in Hoy Cycles.

*(b) Coast between Bay of the Tongue [206 048] and Kame of Hoy [198 048]. Cliffs of Hoy lava and tuff resting on Upper Stromness Flags. Sill of bostonite near sea level (not accessible). Terminal moraine at mouth of corrie [203 043] (see Plate XV).

(c) North coast at Sea Geo [264 025]. Rousay Flags overlain by soft, sulphur-yellow Lower Eday Sandstone.

7. *Hoy II*

*(a) Rackwick and Old Man of Hoy. Coast from Rackwick beach [203 988] (views of Upper Old Red Sandstone Cliffs of Whitefowl Hill) to Too of the Head [194 989]. Lower Eday Sandstone, Hoy tuff and basalts. Retrace steps and join footpath over south shoulder of Moor Fea to coast opposite Old Man of Hoy [177 007]. Hoy Sandstone stack on plinth of lava, resting on Upper Stromness Flags. Return *via* Loch of Grutfea to Ford of Hoy with views of glacially moulded topography.

(b) *Dwarfie Stane* [243 044] and *Dwarfie Hamars*; escarpment of Hoy Sandstone.

Alternative localities worth visiting on Hoy: (a) Coast between Melberry

[266 887] and Ha Wick [250 889]. Amygdaloidal Hoy Lava overlain by cross-bedded, pebbly Hoy Sandstone. (b) South Walls coast, especially Hesti Geo [337 890] and south shore of Kirk Hope [341 895]. Sedimentary structures in Rousay Flags, including fine sun cracks and syneresis cracks. Camptonite dykes.

8. *Rousay*

(a) South Coast at Viera Lodge [392 281] and Fishing Geo [366 310]. Rousay Flagstone cycles with fish and 'Estheria' beds. (Mid Howe chambered cairn and broch are close to Fishing Geo.)
(b) North-west coast between Quoy Geo and Sacquoy Head [382 349]. Sandstones in Rousay sequence, fish beds, high-level storm beach.

9. *Eday I*

(a) *Fersness Bay*, from [532 338] to [544 333]. Complete section from Rousay Flags to high horizon in Middle Eday Sandstone, with very thin sequence of Eday Flags. Freestone quarry in Lower Eday Sandstone and fish bed in Middle Eday Sandstone.
*(b) South-west coast from Point of Sandybank [437 313] to Neven Point [546 297]. Traverse across axial trace of Eday Syncline with unbroken section of Lower Eday Sandstone, Eday Flags and Middle Eday Sandstone.

10. *Eday II*

(a) *South east coast*—from [569 290] to [579 297]. Strata originally taken as Upper Eday Sandstone, now thought to be Middle Eday Sandstone.
*(b) *West coast* from [553 365] to [551 373]. Eday Marl and Upper Eday Sandstone. Fairly accessible sections of Upper Eday Sandstone are also seen near Little Noup Head [560 400].

Note: Good but less complete sections of the Eday Beds are also seen in the south-western peninsula of Sanday. Excellent sections of Rousay Beds are seen in north and west Westray.

B. Shetland Mainland

1. *Lerwick District*

*(a) Coast between Easter Rova Head [474 454] and Scottle Holm [471 449]. Rova Head Conglomerate, folded siltstone and shale, dykes and ?vents of tuffisitic breccia.
(b) South Ness [483 405] to Horse of the Nabb [479 403]. Sedimentary structures and rhythmic sequences in Lerwick Sandstone, some large plant remains. Irregular veins of pink calcite at the Nabb.
(c) Ness of Sound, coast from [468 384] to [472 391]. Two lacustrine limestones interbedded with Lerwick Sandstone facies.
(d) Gulber Wick, north shore from [446 386] to [450 385]. Typical sediments of Lerwick Sandstone facies. Also (if time available), south shore, at Ness of Setter and Fea Geo [447 370]—sediments transitional to Brindister Flagstone facies.
*(e) Coastal sector north of Bay of Fladdabister at [439 327] or [443 337]. Basal breccia resting on undulating metamorphic basement. (Good

sections of basal breccia are also seen at and near the north shore of East Voe of Quarff [434 354].

2. *South-east Shetland: Old Red Sandstone*

*(a) Scatsness Peninsula and Ness of Burgi [389 087]. Extensive sections of conglomerates, pebbly sandstones and trough-cross-bedded sandstones, red siltstones and a fish bed.

(b) Sumburgh Head (i) The Slithers seen from [409 084], lacustrine calcareous flagstones affected by landslips. (ii) Roadside Quarry [406 093] ?lacustrine siltstone with minor sedimentary structures. (iii) Loos Laward [410 103] small- and large-scale slumping in lacustrine calcareous flags.

*(c) Coast between The Cletts [407 127] and Vaakel Craigs [410 136]. Exnaboe Fish Bed, fluvial conglomerates and sandstones with pebbletrains, and ?aeolian sandstones with large-scale cross-bedding.

3. *Old Red Sandstone of Walls Peninsula*

*(a) Coast between Wick of Watsness and Voe of Footabrough—(best section between [185 496] and [193 493]). Folded sediments of Walls Formation with rhythmic sedimentary sequences. Well-developed cleavage, lineation and small-scale plastic folds in fine-grained sediments. Very rare fish and plant remains in some siltstones.

(b) Coast of Wats Ness between [173 503] and [173 590]. Non-tectonised sequences of Walls Formation, accessible at Trea Wick—(rare fish and plant remains in fine sediments).

(c) Voe of Dale [174 522] and coast from Mu Ness [165 524] northwards. Sediments, pyroclastic deposits and felsite sill in Sandness Formation. Fine views of spectacular cliffs and coastal exposures of Sandness Formation sediments looking north from Mu Ness.

(d) Clousta area (i) North shore of Voe of Clousta [306 575] to [296 581]. Sedimentary sequence with basalt flow and interbedded cone of basaltic pyroclastics. (ii) Exposures north of road [300 572]—Acid tuff and agglomerate with ignimbrite clasts.

4. *Metamorphic Rocks and Old Red Sandstone, Walls Peninsula*

*(a) Sandness Coast from Bay of Garth [214 581] to Skerry of Stools [221 592]. Schists, quartzites and limestones with D2 lineation and boudinage and minor folding, locally re-folded by D4 conjugate folds. Porphyrite dyke.

(b) Melby Coast. *(i) Shore between [172 572] and [180 576] Lower Melby Fish Bed and associated strata. (ii) Shore at Pobie Skeo [168 567] Upper Melby Fish Bed and associated strata. (iii) Ness of Melby [185 850] Rhyolites and underlying sediments.

5. *Old Red Sandstone lavas and pyroclastics of Esha Ness*

*(a) Coast south of Esha Ness Lighthouse [205 785]. Cliffs of andesitic lavas and tuffs with caves and blowholes.

(b) Coast between Stenness [213 773] and Fiorda Taing [227 768]. Andesitic lavas and tuffs; viewpoint for Dore Holm natural arch.

*(c) Grind of the Navir [213 805] and coast as far south as [213 795].

Ignimbrite, high-level storm beach, mugearite cliffs, subterranean passage and 'gloup'.
(d) Heads of Grocken [264 774]. Cliffs of Ronas Hill Granite and junction with Hillswick metamorphic rocks: dykes.

6. *Fethaland and Northmaven*

(a) Circuit of Fethaland peninsula (a full day) or east coast only. Special features: Old Red Sandstone outliers, e.g. at [376 915], talc with magnetite octahedra [376 936], large garnets in schist [375 944], serpentinite associated with talc and tremolite [377 496].
*(b) Exposures near road: (i) Hillside west of Laurence Loch [360 859]. Screen of metamorphic rock along granite margin. (ii) The Brig [356 844]. Eastern granite ring dyke, serpentine, etc. (iii) [353 834]. Striped gneisses. (iv) [337 807] Ultrabasic plutonic rock bounded by metamorphics. (v) Old magnetite mine, Clothister Hill. [342 729]. Magnetite, garnetiferous skarn rocks. (vi) Roadside Quarry at [342 682]. Scapolite-calcite veins in diorite.
(c) Coast south of Gunnister Voe, between [300 725] and [305 700]. A specialist igneous excursion. Extensive exposures of granite-diorite-basalt net-vein complex. Access from road end at Nibon. [305 731]. (This ½ day+ excursion can be substituted for 6(a) or (b) or 5(b) and (d).

7. *'Lewisian' Orthogneisses, North Roe.* (10 mile walk) Coastal exposures of continuous interest from [365 915] to [306 901]. Also interglacial peat at Fugla Ness [312 913]. Return by Beorgs of Uyea: riebeckite-felsite dykes, neolithic quarry and workshop at [327 901].

8. *Lax Firth Area*

(a) Coast between Fora Ness [461 475] and Hawks Ness [462 491], exposing Dales Voe Grit and Clift Hills Phyllite. Tight folding with axial plane schistosity and lineations, also large-scale boudinage of epidiorite masses in phyllite. Continue south to [450 480], Asta Spilitic Group. Access from road end at Breiwick [455 474].
(b) Wadbister Ness. The section commences at [446 503] in the Girlsta Limestone, followed to the east by the Wadbister Ness Group. The remainder of the Wadbister Ness Group and the Laxfirth Limestone can be seen by continuing southwards along the shore of Lax Firth. Access from road end at Wadbister [435 495]. (For alternative see 11(e) and (f).)

9.*(a) *Scalloway area*

Shore section from Westshore [397 389] to Ness of Burwick [388 407], exposing migmatised and non-migmatised rocks of the Colla Firth Group including the East Burra Pelite. Numerous intrusions of foliated granite and pegmatite associated with the migmatitic gneisses; also a few dykes of porphyritic microgranite and microdiorite which possess a schistosity but were intruded after the migmatisation.

(b) *Cunningsburgh area*
(i) Shore section from North Voxter [438 287] to Croo Taing [429 269].

Metamorphosed basic lavas, agglomerates, tuffs and intercalated meta-sediments, mainly graphitic phyllite.

(ii) Traverse across hillside from [424 271] to [423 284]. Talc-magnesite rock with residual masses of serpentinite; also metamorphosed basic lavas, some with original porphyritic texture well preserved, and epidiorite. Typical Dunrossness Phyllite with chloritoid and quartz segregations at north end of traverse.

10. *Eastern Plutonic Complexes*

  *(a) Coast between Firth [440 733] and Mossbank [450 757]. 'Inclusion Granite' composed of varied suite of plutonic rocks with hornblende xenoliths and enclaves of country rock.

  (b) West shore of Sullom Voe at Ness of Bardister [370 760] and Gluss Isle [370 780]. Spectacular exposures of 'Inclusion Granite' and pegmatites.

  *(c) North shore of Olna Firth and east shore of Busta Voe, between Wetherstaness [357 652] and Saltness [362 668]. Brae Complex.

  (d) Coast opposite Fora Ness [442 723]. Fine view-point of treble ayre linking Fora Ness to Mainland; peat banks submerged at high tide.

11. *Lunnasting and Nesting*

  *(a) Roadside and hills around [497 667]. Pelitic gneisses with crystals of kyanite, sillimanite, staurolite and tourmaline (at 500 664]).

  (b) Coast of Lunning Head from [510 668] to [506 674]. Coarse gneisses of Scatsta Quartzitic Group. Serpentine at [506 673].

  *(c) Lunna Ness, east coast between [494 694] and [476 672], and inland exposures around [495 697]. Coarse augen-gneisses with large feldspar porphyroblasts.

  (d) Coast at Boady Geo [472 560]. Permeated gneisses of Colla Firth Group with coarse pegmatites.

  (e) Coast of South Nesting between [453 530] and [459 530]. Girlsta Limestone.

  (f) Coast at Gletness between [463 513] and [470 514]. Good section in Wadbister Ness Group.

C Shetland: Smaller Islands

For visitors who wish to extend the excursion to include the islands readily accessible from Mainland the following itineraries are suggested:

Unst I

  (a) *Valla Field Block*. Traverse block from Balliasta [600 096] westward to Hagdales Ness [574 092], thence south along coast to Collaster [578 078], returning via Berry Knowe [586 074]. Lithostratigraphic units of Valla Field Block. Along northern traverse and on northern part of shore section the rocks have the mineral assemblage developed during the First (prograde) Metamorphism. Farther south the minerals formed by the Second (retrograde) Metamorphism are overprinted.

  *(b) *Main Serpentine Block*. Nikka Vord [624 105] to Muckle Heog [631 108]. Peridotite serpentinite with old chromite quarries, where

specimens of chromite, kämmererite, zaratite, etc. can be collected. Continue *via* Hagdale [638 108] (talc belts) to Keen of Hamar [646 097]. Upward passage of peridotite-serpentine into pyroxenite.

Unst II

*(a) Burra Firth, south-east shore [618 142]. Deformed and folded rocks of Valla Field and Saxa Vord blocks; deformed Queyhouse Flags.

*(b) Queyhouse Quarry [611 124]. Fine exposures of talc and antigorite-serpentine along sheared western margin of Main Serpentine Block.

(c) East slope of Housi Field [632 140]. Blocks of Saxa Vord schist with large kyanite crystals.

(d) North shore between Virdik [653 174] and Forn Geo [663 170]. Skaw Granite with enclaves of gneiss.

*(e) Nor Wick from [654 152] to [653 142]. Shear zone at base of Skaw Granite with 'woody' structure in schists. Graphite- and hornblende-schists of Nor Wick Hornblendic Series affected by Third (dislocation) Metamorphism.

(f) Clibberswick [651 121]. Talc at base of Clibberswick serpentinite; remains of ancient soapstone workings.

Unst III

*(a) Coast section from Uyeasound [590 010] *via* Wick of Belmont to Snarra Voe [565 018]. Effects of dislocation metamorphism on meta-gabbro, serpentine and schist. Shear belt at base of Lower Nappe at Belmont just south of jetty [565 005]. At Winner Houll [557 009] and Snarra Voe [565 018] chloritoid-gneisses of Second Metamorphism with relicts of staurolite, kyanite and garnet.

*(b) *Mu Ness.* Ness of Ramnageo [624 000] to [630 001]. Mu Ness meta-gabbro and serpentine, Muness phyllites with deformed conglomerates, spessartite dykes and sills.

Note: For a more detailed and extensive itinerary of Unst see Read (1936b).

Fetlar

*(a) *Funzie Conglomerate.* (i) Coast from Stava Ness [654 887] to The Snap [657 878]. (ii) Coast at The Tind [673 904].

*(b) *Wick of Gruting.* [660 910] to [644 920]. Serpentine thrust over hornblendic and graphitic schists, chloritoid-phyllites and greenschists; some deformed micro-conglomerates.

(c) *Wick of Tresta.* Coast between [608 905] and [619 902]. Lineated graphitic and hornblendic schists and some serpentines.

(d) North coast from [605 936] to [612 940]. Schuppen-zone with deformed phyllonitised conglomerates overthrust by Saxa Vord Serpentine.

Bressay

*(a) Bight of Ham between [492 402] and [487 398]. Sediment-breccia with enclaves of vertical and inverted sandstone.

*(b) Muckle Hell Vent [526 400]. Vent of sediment-breccia with dykes of intrusive carbonate rock.

(c) North-west coast between Scarfi Taing [476 445] and Ness of Beosetter [493 445]. Conglomerates, pebbly sandstones, flagstones and sediment breccias with carbonate dykes and veins.

Papa Stour

(a) Coast between Hirdie Geo [151 605] and Aesha Head [147 611]. Basalt lavas with amygdales; Lower Tuff truncated by Lower Rhyolite.

(b) Geo of Bordie [155 624]. Middle Tuff on irregular eroded base of rhyolite (see Plate VIIIB).

(c) Sholma Wick [161 619]. Middle and Lower Tuffs and agglomerate; highly irregular top of Lower Rhyolite.

(e) Coast between Jerome Coutt's Head [170 622] and Willie's Taing [184 622]. Unweathered top of Lower Rhyolite with 'pillow' structure; Middle Tuff.

INDEX

144

Printed in Scotland for Her Majesty's Stationery Office
by Bell & Bain Ltd., Glasgow. Dd.020046/3492 K160 4/76.